Innovative Economic Symposium (IES2022)

Opportunities in Post-Covid Era

SHS Web of Conferences Volume 160 (2023)

Setubal, Portugal
9-10 November 2022

Editors:

Jakub Horák
Ana Rolo
Eva Dolejšová

ISBN: 978-1-7138-7099-9

Printed from e-media with permission by:

Curran Associates, Inc.
57 Morehouse Lane
Red Hook, NY 12571

Some format issues inherent in the e-media version may also appear in this print version.

No changes have been made to the content of these proceedings. There may be changes to pagination and minor adjustments for aesthetics.

Printed with permission by Curran Associates, Inc. (2023)

For additional information, please contact EDP Sciences – Web of Conferences at the address below.

EDP Sciences – Web of Conferences
17, Avenue du Hoggar
Parc d'Activité de Courtabœuf
BP 112
F-91944 Les Ulis Cedex A
France

Phone: +33 (0) 1 69 18 75 75
Fax: +33 (0) 1 69 28 84 91

contact-edps@webofconferences.org

Additional copies of this publication are available from:

Curran Associates, Inc.
57 Morehouse Lane
Red Hook, NY 12571 USA
Phone: 845-758-0400
Fax: 845-758-2633
Email: curran@proceedings.com
Web: www.proceedings.com

TABLE OF CONTENTS

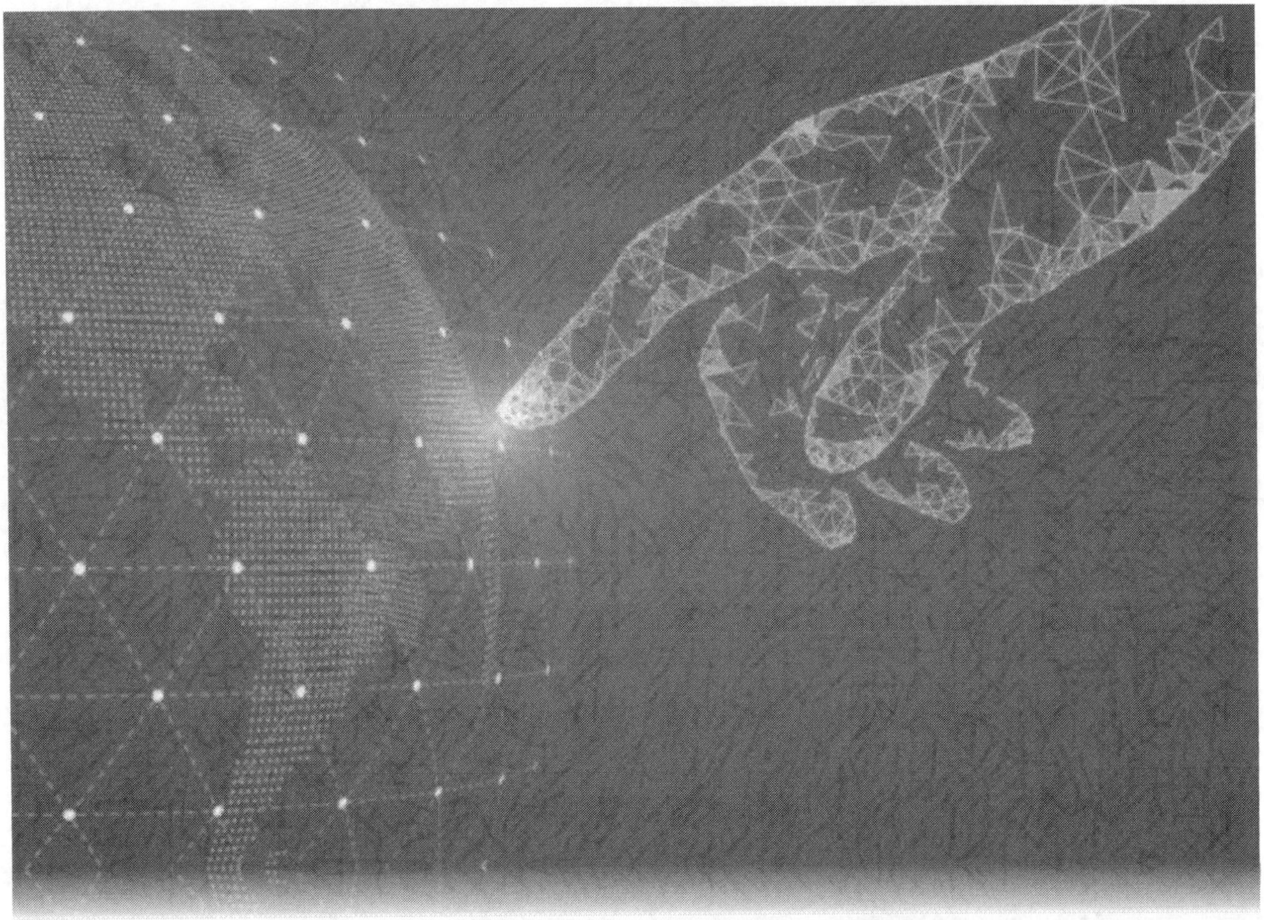

Innovative Economic Symposium 2022

Opportunities in post-covid era

IPS Instituto Politécnico de Setúbal
Escola Superior de Ciências Empresariais

VŠTE

Innovative Economic Symposium 2022

Opportunities in post-covid era

http://ies-conf.com/

Editors: Jakub Horák, Ana Rolo, Eva Dolejšová

About the conference IES 2022

Date of the conference:

2022-11-09

Motto

"Coming together is a beginning; keeping together is progress; working together is success."

Henry Ford

Target

The IES 2022 conference brought together scientists and practitioners in economy, international relations, finance and marketing from allover the world. It aimed to collect innovative ideas about new trends in business and corporate finance in the current situation of the world economy.

Thematic Areas

- Reverse logistics, sustainability, recycling
- Sustainability: new pathways of learning & teaching and industry involvement
- Economic impacts on business, automation
- Use of artificial intelligence in business
- Micro & macroeconomic impact on economic indicators and GDP
- Impact on the banking sector, international business
- Influence on stocks, commodities, circular economy
- Labor markets and Human Resources Management
- Digital marketing and costumer behavior
- Financial Reporting and the role of accounting
- Information Systems
- Logistics
- New technologies across management

About organizers

Polytechnic Institute Setúbal, Portugal

IPS (Polytechnic Institute of Setúbal, Portugal) is a public institution of higher education situated in the region of Setúbal (45 km away from Lisbon). IPS was created in 1979 and started activity in 1981. At present the IPS comprises five Higher Education Schools.
The IPS has two campi: the Setúbal campus, where four Schools are located - Setúbal School of Technology, School of Education, School of Business and Administration, School of Health - and the Barreiro campus, where the Barreiro School of Technology is located, in a total of approximately 6500 students and 650 employs.
ll the premises and equipment are high quality and up–to–date. There are a number of well-equipped laboratories, documentation centers, audio-visual equipment, and computer facilities available to students and teachers.

More information at https://www.ips.pt

Institute of Technology and Business in České Budějovice (VŠTE)

The primary focus of VŠTE activities is the implementation of full-time and part-time studies for professional bachelor's and master's degree programmes. The professional study programmes are supported by cooperation with regional entrepreneurs.
VŠTE´s other priorities include applied research, the implementation of innovative technologies and knowledge transfer into areas that are current and crucial with regards to the regions needs. Scientific working groups focus on variety of research targets across economicla nad technical fields, incl. artificial intelligence, neuron webs, corprate strategy, corprate finance, valuation, and many others.

More information about VŠTE at www.vstecb.cz.

Conference report

In the light of recent covid measures, this year, the conference took place online again using the communication platform ZOOM. The whole symposium was organized by the team of organizers from Polytechnic Institute of Setúbal, Portugal and from the Institute of Technology and Business, the Czech Republic.
On 9th November 2022, academics from Ukraine, Portugal, Poland, Slovakia, South Korea and the Czech Republic joined the virtual meeting to present their research and exchange expert knowledge.

This year, the thematic focus was "Opportunities in post-covid era" which allowed describing the opportunities and changes that raised after the covid period.
The conference started with opening speech of A/Prof. Ing. Jarmila Straková, Ph.D Vice-rector of the Institute of Technology and Business in České Budějovice (VŠTE) and A/Prof Luísa Carvalho Ph.D. Vice President of Polytechnic Institute of Setúbal, Portugal, the main host of this year's conference. The conference programme consisted of partly of one and partly of two parallel plenary sessions. All presented contributions attracted the attention of wide audience. Each participant received The Book of Abstracts containing brief descriptions of the presented contributions.

Expert guarantor

prof. Marek Vochozka - Former rector of the Institute of Technology and Business, Czech Republic

Pedro Pardal, MSc, PhD. - Dpt. of Accountancy, Polytechnic Institute of Setúbal, Portugal

The Organizing Committee

The organizers in the host country (Portugal):

Ana Rolo - Polytechnic Institute of Setúbal
Pedro Pardal - Dean of the Polytechnic Institute of Setúbal
Paulo Bogas - Polytechnic Institute of Setúbal
Rui Alves - Polytechnic Institute of Setúbal

Organizers in the Czech Republic:

Eva Dolejšová, MS, PhD - ITB in Ceske Budejovice (CZE)
Jakub Horák, MSc, MBA - ITB in Ceske Budejovice (CZE)

The Scientific Committee

Statement of Peer review

In submitting conference proceedings to SHS Web of Conferences, I certify to the Publisher that I adhere to the **Policy on Publishing Integrity** of the journal in order to safeguard good scientific practice in publishing.

1. All articles have been subjected to peer review administered by the proceedings editors.
2. Reviews have been conducted by expert referees, who have been requested to provide unbiased and constructive comments aimed, whenever possible, at improving the work.
3. Proceedings editors have taken all reasonable steps to ensure the quality of the materials they publish and their decision to accept or reject a paper for publication has been basedonly on the merits of the work and the relevance to the journal.

Title, date and place of the conference

Innovative Economic Symposium 2022, Opportunities in post-covid era.
November 9 - 10, 2022, Setúbal, Portugal

Proceedings editor(s):

Jakub Horák, Ana Rolo, Eva Dolejšová

Date and editor's signature

17, avenue du Hoggar - PA de Courtabœuf – BP 112 - 91944 Les Ulis Cedex A (France)
Tél. : 33 (0)1 69 18 75 75 - Fax : 33(0)1 69 07 45 17 – www.edpsciences.org

Co-creation design thinking process: Learning with demola approach

Célia Picoito[1], *Helena* Caria[1], *Fernando* Angelino[1] *Alcina* Dourado[1], *Nuno* Pereira[1], *Fátima* Serralha[1], *Alice* Ruivo[1], *João* Vinagre[1], *Bernardo* Ramos[2], and *José* Pires[3]

[1]Instituto Politécnico de Setubal, Campus do IPS – Estefanilha, 2910-761 Setúbal, Portugal
[2]Escola de Hotelaria e Turismo de Setúbal – Av. Luisa Todi nº 5, 2900-461 Setúbal, Portugal, Largo da Escola Vila Nova de Santo André, 7500-160 Santo André, Portugal
[3]Escola Secundária Padre António Macedo em Santo André - Largo da Escola Vila Nova de Santo André, 7500-160 Santo André, Portugal

Abstract. Nowadays we witness numerous partnerships and projects of national and international scope around active training, using digital tools and based on co-creation processes between different partners, to obtain more creative and appropriate processes to respond to current and future scientific, academic and professional challenges. This article presents the participation of a group of professors from Instituto Politécnico de Setúbal (IPS) in the DEMOLA® approach to be trained as facilitators. We present the objectives, the formation of the teams and the established work methodology, and the application of this methodology adapted to the curricular unit of Process Optimization of the Master of Biological and Chemical Engineering.
We briefly describe the main phases that constitute the development of the project within the scope of IPS and the obtained results from the analysis of all components (student training, professors, challenges with companies/organizations and perceptions of the whole team).
We emphasize the development of instrumental skills (Canva, Miro, PESTLE, Portfolio); scientific/ pedagogical skills (Teamwork, Multi-professional Approach, Co-creation Processes, Design Thinking, Research, Facilitation, Disruptive Thinking and Co-construction, among others); and "win-win" situations, or rather "learn-learn" between facilitators and students, innovation, intergenerational, sustainability and freedom of thought.
Keywords: Demola approach, Co-creation, Learning, Design Thinking

1 Background

DEMOLA® is a co-creation approach, created in 2008 within the innovation ecosystem from Tampere, Finland, which involves students and external organizations intending to generate ideas based on challenges. It has in its origin the collaboration of the country, local universities and the private sector [1].

The DEMOLA® platform (www.demola.net) facilitates local and international co-creation projects between students, companies, and higher education institutions. DEMOLA® is a co-creation approach focused on solving real challenges. Each project carried out will have a positive result – be it a new concept, a demonstration, or a prototype. If the partner company finds the result useful, it can license or buy the result and promote its further development. Each partner has a clear role, and the work is guided by simple procedures. Contracts, intellectual property rights, licensing models and other legal requirements are in place and compliant with international business standards and practices [2].

For companies, the co-creation approach helps to explore future impacts and better understand the requirements for their products or services, exploring changes in consumer behaviour. For students, it is an enriching experience during their studies, providing close contact with companies, non-governmental organizations (NGOs) and public sector organizations. For professors, the aim is to increase collaboration between academia and industry and establish a joint working platform between the various institutions participating in the DEMOLA® platform at the national level [3]).

The value of co-creation generated through the DEMOLA® methodology, based on a valid strategic approach to social challenges, has become more widely recognized, which has led to an increased interest in testing it with new multi-sectoral and multi-stakeholder approaches and may even help to overcome some of the university cultural tensions [4]

Co-creation is broadly defined in the literature [5, 6] as the creation of value by participants in the process and activities of the value chain, through their active participation. In this process, high-quality interactions [7, 8] allow an individual customer to co-create unique experiences with the company, which may unlock new forms of competitive advantage [9]

The Organization for Economic Co-operation and Development (OECD) considers a co-creation process as the interaction between all actors through the free circulation of knowledge and transformation into products and services, which support the growth of new markets, entrepreneurship, and the promotion of an entrepreneurial culture [10]. Citizen participation in co-creation and innovation processes is one of the key issues in terms of European Union policies, which highlights some needs such as: having an objective and a clear method of application; sharing responsibilities; being transparent; having enough time; be open. Sometimes, the most important thing is not the result of the co-creation process, but the social change that begins after this cooperation, which helps to gather information about existing social issues in the community and to identify areas where to develop more innovative practices, making them more inclusive and participatory co-creation process [2, 10] In search of solutions for a better future, the EU policy agenda for 2030 promotes actions that stimulate the co-creation of innovations, sustainability goals and Sustainable Development Goals [11]). In this sense, one should seek to understand the most complex challenges through the integration of different approaches in a sustainable development system and interactions between various disciplines, promoting co-creation between the various local actors and actors of the scientific and technological system [11]. The next generation of Science, Technology, and Innovation (STI) policies will have as fundamental principles a clear orientation of their mission and co-creation processes and will be implemented through dynamic public-private partnerships [4].

Companies and industry in general, are aware of the relevance of co-creation processes within the Design Thinking approaches for the successful development of new products or services. This clarifies the relevance of co-creation methods for the acquisition of the now-named future skills applied to teamwork, to master the challenges of the future namely communication, empathy and conflict management, cooperation, ethical and decision competencies and also facilitation methods, identity and cohesion of teamwork [9]).

We can highlight that the success of a co-creative design-thinking process is directly articulated with the facilitation process. The facilitator has the role of being a coach that guides and stimulates the team participants to evolve their future skills and to promote the best interaction between participants.

There are a variety of personal qualities deemed necessary for team facilitators [12, 13]), including self-awareness, self-confidence, concern and respect for others, non-judgmental acceptance, genuineness, empathy, vitality and maturity. Simply, the role of the team facilitator is to create an atmosphere of psychological safety that promotes self-disclosure, feedback and experimentation associated with alternative behaviours, contributing to an attitude of constant curiosity like the one in the basis of scientific research, however not considering the limits of known possibilities to think out of the box.

There is also a strong link between empowerment and the facilitation processes to support local partners and other stakeholders, first by defining challenges and identifying needs prioritization, and second by developing solutions and making decisions that could be further accepted and integrated into the community [14]. Participatory decision-making processes emerge from this approach as being based on four fundamental values [15]: full participation; mutual understanding; inclusive solutions and, shared responsibility. Also, according to the same authors, meetings conducted according to these participatory values are described to produce significant results, contributing to strengthening individuals, teams and agreements, as examples of many other benefits.

Individuals and groups empowered in this way will be better equipped for future challenges of the job market, exhibiting more skills and leading to a supportive atmosphere generated within the team. Agreements reached in these conditions will include higher-quality ideas, wiser goals and more inclusive solutions [15, 16].

DEMOLA approach is a strong commitment to innovation, multidisciplinary, and internationality. Simultaneously it is also committed to developing a methodology to improve a wide range of skills in students and professors of higher education institutions (HEIs) that will achieve facilitators skills. DEMOLA's approach has the ambition to strengthen collaborative networks of local and global companies and HIEs professors, bringing this collaboration to the learning environment.

DEMOLA methodology equips the teams with a set of tools for co-creation, providing the acquisition of strategic forecasting work and its integration as teaching tools. The remote work methodologies, nowadays a common approach, allow the creation of international and multidisciplinary teams, sharing different visions, cultures and attitudes, which enrich the search for innovative solutions.

The main objective of the DEMOLA approach is to create a culture and operational practices that will enforce collaborative networks between national polytechnic institutions as well as the transfer between industry and academia. The relevance of these networks will bring new ideas to the class, highlight research niches and, consequently, create opportunities for new research domains. From an Educational and Pedagogical perspective, it improves teaching practices; brings actual and relevant content to teaching and contributes to the development of future skills, increases also professional identity and employability rates at the student's level.

DEMOLA approach is based on co-creation design thinking processes, thus is based on teamwork practices enabling the participants to:

- identify real challenges, arising from the day-to-day stakeholder's context, namely industry or companies or public and social institutions;
- work in multidisciplinary teams of students, led by facilitators (professors from HEIs);
- look for innovative solutions to the challenges identified.

2 Objectives

The main goal of this paper is to share the experience of a group of IPS teachers that are DEMOLA facilitators after performing a 15 week of training according to the DEMOLA® approach. The share of this experience and its application to a classical unit from an IPS Master program as a follow-up of the new skills achieved are also presented. The DEMOLA training of facilitators has the objective of developing new skills in co-creation design thinking processes that could be applied to teaching nowadays.

3 Methodology

The DEMOLA approach in Portugal, and specifically at IPS, was actively developed for 15 weeks, each one dedicated to different stages of the co-creation process. A previous stage was the selection of HEIs professors to join the project.

The 15 weeks of training span the 1st semester of the academic year of 2021-2022, running between September and February.

In the first six weeks, professors were trained to submerse in the approach and develop skills as facilitators. At this stage, some of the specific tasks undergone were the identification of a company to become the partner entity, the identification of a challenge to be publicized to attract students, since the student's enrolments occur during this step, and the signature of confidentiality, ethical and legal agreements.

The second stage is composed of eight weeks of teamwork, divided into two phases: the analysis of the challenging context following a PESTLE analysis; and the identification of future perspectives.

1. The DEMOLA process and the enrolment of companies' partners
2. Definition of a challenge
3. Enrolment, selection and creation of the student team
4. Online tools and Teambuilding
5. The skills of a facilitator
6. Project start-up: Discovery phase
7. Project kick-off of the student team (week 1/8)
8. Building trust (week 2/8)
9. Work phase (week 3/8)
10. Facilitation skills with the project partner (week 4/8)
11. Work phase (week 5/8)
12. Decision-making support, solution refinement and finalization (week 6/8)
13. Preparation of the presentation meeting (week 7/8)
14. Final project meeting (week 8/8)
15. Portfolio reflection of the work performed.

Of all these phases, we feel the need to highlight professors' training, which leads them to abandon their traditional role of educator-trainer to play the new role of facilitator.

Students were also subjected to a training process, capable of enabling them to carry out PESTLE analysis and to use collaborative work development tools (e.g. MIRO). These two tools are considered essential: PESTLE analysis allows the users to build knowledge of the reality of the phenomenon under study. It focuses on external forces that the company faces, and the letters stand for Political, Economic, Social, Technological, Environmental and Legal – PESTLE. The MIRO tool narrows the distance between the participants by allowing them to complement the communication between those involved. It should be noted that, whenever possible, it is desirable to establish a relationship based on the physical presence of students and facilitators, which happened in several cases, including activities such as visits to the companies.

It is also important to stress the holding of two bootcamps in which IPS professors socialized and exchanged ideas with professors from other polytechnic institutions in the country, who were also facilitators of this DEMOLA approach. This enabled the sharing of knowledge acquired and the best techniques to pursue the objectives of the different projects.

The 10 IPS professors, worked in pairs as team facilitators and developed 5 projects during the 15 weeks of training.

The evaluation of the DEMOLA project is made by considering different aspects: one directed towards the students and the other towards the facilitators. There was also a global assessment of the proposed solutions to the problems presented (carried out by each team) and an overall assessment of the methodology used.

The students prepared two reports:

- one in the middle of the process, with contextualization and deepening of the theme in the light of the present, and materialized in the PESTLE analysis;
- a final report, where the solutions found were presented and justified.

For facilitators, the experience culminated with the elaboration of a personal portfolio, where each one exposed their personal views regarding their participation in the project, the difficulties experienced and the learning achieved, thus enabling a good reflection on the methodology applied.

There were also meetings between each team and the partner organizations/entities, to present the proposed solutions. The discussion and satisfaction of the same with the work developed should be highlighted in the facilitator's feedback to the students.

4 Results and Discussion

A global assessment of the experience in the DEMOLA® approach allows us to analyse the contribution of facilitators and students to develop new skills.

The major contribution of the DEMOLA® approach is based on networking through digital media of facilitators and students.

Major results from the DEMOLA® approach can be indicated as tools for the facilitating process:

Instrumental skills, by using Canva, Miro, PESTLE analysis and Portfolio;

Scientific/pedagogical skills using: teamwork, multidisciplinary approach, co-creation processes, design thinking, research, facilitation, disruptive thinking and co-construction, among others;

The ability to analyze a reality and present solutions to real problems, through the development of future skills, the enhancement of students' skills, the contention that facilitators had to train in order not to present solutions to students, a "win-win" situation, or rather “learn-learn” between facilitators and students, innovation, intergenerational relations, sustainability ideas and freedom of thought - different from the usual, with no known solutions –think out of the box!

From the perspective of the facilitators, also IPS teachers, the use of these results contributes to a shift from education in the classical sense, towards direct support for the practice in real context by allowing to establish new formats such as mentoring, coaching, reflection support, networking or formation of learning communities.

Also relevant as results are the 5 projects developed by the 5 teams with IPS facilitators. The main guidelines for each project are presented in table 1 and were established according to the challenges presented by each team stakeholder.

The share of these challenges, developed by each team, in co-creation with facilitators, students and stakeholders allow us to illustrate how the co-creation design thinking approach can be used in innovative domains far from the professional formation of each facilitator. Co-creation design thinking approach allows the teams to act as proactive global citizens who help design the environment and solve problems.

Table 1. Summary of the 5 projects developed by the Teams with IPS facilitators of the 1st semester of 2021-2022.

DESIGN OF A NEW WORLD… WITH PERMACULTURE (Alcina Dourado and Bernardo Ramos)	
Partner	Exceptionally, this challenge did not imply a company but involved a project as an expert. The promoters of the Project work in the field of permaculture and are based in central Portugal.
Team	International Master's student in Psychology (University of Pavia, Italy) and a nutritionist student of Biochemistry and Health, in the field of Clinical and Metabolic Biochemistry (Polytechnic Institute of Porto).
Challenge	With climate change, the future of humanity is at risk. It is essential to understand the contribution of ecologically based movements to a more responsible, sustainable and, above all, regenerative life. How to make it accessible and attractive to the masses? Permaculture is the design methodology for regenerative human lifestyle systems. Human habitats, work with Nature, not against Nature. It works as an anchor that connects the citizen to essential principles, techniques, and methods. It seeks food sovereignty, land management, socio-economic well-being, and ecological regeneration responsibly, defending environmental sustainability and leading to a new reality. However, it faces scepticism, various stereotypes, and prejudices. In other words, there is a lack of evidence that a permaculture is a viable option for the restructuring of conventional agriculture, being accused of proposing a utopian (and therefore) impossible world. How can we strengthen the conditions to create a resilient world by making permaculture mainstream? How would it be?

GENZ GREENOWER: CIRCULAR ECONOMY (Alice Ruivo and Célia Picoito)	
Partner	A technological and innovative company, specialising in intralogistics, that is, in the search for continuous improvement, increasing the competitiveness of its customers through the implementation of specific solutions as an integral part of its performance.
Team	Four students, two from Accounting and Finance (ESCE/IPS), one from Marketing (ESCE/IPS) and one from Sea and Environmental Technology (ESTSetúbal/IPS), all from the Polytechnic Institute of Setúbal.
Challenge	In industry, waste accumulates and involves expenses related to treatment or pollution. Materials for new products must come from old ones – the aim is to reuse as much as possible, reproduce, recycle raw material, or use it as a source of energy. The partner company has waste as a result of a production process - clean plastic from cutting plastic materials - and has not yet found a way to reuse it. We challenge Generation Z to understand this phenomenon, think about it and propose a solution. It can be a new product for the company, or to be used by another company, a source of energy... the limit is your imagination. You can put into practice what you need to change: stop wasting! Start the circular economy! And you can be an active part of this transformation! Do you accept the challenge of showing how Gen Z has a Greenpower?

ROAD MAP TO A CARBON-NEUTRAL CITY (Fernando Angelino and João Vinagre)	
Partner	Regional Energy Agency of the municipalities of the Lisbon and Tagus Valley region. A private non-profit association, created with the support of the "Intelligent Energy-Europe" program and the Municipal Councils in its area of intervention.
Team	Two students from ESTBarreiro/IPS (Bachelor in OilTechnologies), one student from ESS/IPS (Master in Physiotherapy), and one Brazilian student (Master in Business Sciences).
Challenge	The transition to a climate-neutral society is both an urgent challenge and an opportunity to build a better future for all. How can these problems be solved when applied to a territory such as the Baía do Tejo Business Park, in Barreiro? What strategy should the territory adopt to combat climate change? What decisions should be taken in the field of energy efficiency and renewable energies? Are there effective ways to accelerate the transition to a sustainable industrial and business park? Are there preventive measures to ensure this objective? What effective guidelines should be taken? How should an industrial park be built to meet the targets for net CO2 Zero emissions by 2050? What typology of companies and industries should be attracted to the territory?

THE PATH OF STONES (Nuno Pereira and José Pires)	
Partner	Portuguese cement production company.
Team	Five students from ESTSetúbal/IPS: one from Mechanical Engineering, two from Environmental and Marine Technologies and two from Energy Technologies.
Challenge	Making the social and economic development of a region compatible with the defence of the landscape and the environment is fundamental for harmony, although it is not always easy to achieve It is essential to invest in a strong technological upgrade of the industrial process to improve operating conditions and preserve the environment, but will it be enough to satisfy all stakeholders? How can we find the balance between the natural park and the cement industry that has existed in that location for a hundred years?

SPIRULINA – GREEN FOOD 4 HEALTH (Helena Caria and Fátima Serralha)	
Partner	A small company in the area of artisanal food production, from the central region of Portugal
Team	One student from Mater in Biological and Chemical Engineering and two students from Biotechnology, all from ESTBarreiro/IPS; one student from Biomedical Technology at ESTSetúbal/IPS.
Challenge	Spirulina is a green-coloured cyanobacterium integrated into the microalgae groups. It is currently considered by the world health organization (WHO) as a complete and innovative food being produced in an artisanal and sustainable way this business, gives it healthy properties and a great taste, maintaining its green colour. The challenge aimed to identify how it is possible to: 1. Increase the consumption and acceptance by consumers of green, but very healthy and complete food, and simultaneously, create consciousness of its local and sustainability production; 2. Create communication channels with innovative recipes; 3. Increase the involvement of the Gourmet/Chefs cuisine in this new and natural cooking practice

It is intended by the authors, that this methodology could be replicated in some of IPS's Curricular Units, contributing to the implementation of co-creation design-thinking process practices in the training of future graduates. In this way, the aim is to contribute to the training of new professionals who can establish a real connection between emerging themes, in different contexts, and the creation of solutions outside the usual patterns and based on multi-professional approaches.

With this objective, the Co-creation methodology has already been applied in the curricular unit of process optimization of the master's degree in biological and chemical engineering. A challenge has been proposed on how to contribute to reducing the use of materials and energy. After carrying out an analysis, students proposed an approach that was studied by the PESTLE analysis and presented implementation scenarios.

The following stand out about the students' opinions:

" I liked the methodology because it was necessary to research several points"

"it would be interesting to focus on the solutions"

"It was a good work to learn more information on the topics covered, among which several laws that we did not know of their existence"

" It also allowed us to know the PESTLE analysis, of which we were also not aware"

" the work was interesting and good for my learning because it allowed me to learn how to visualize a particular problem in the various perspectives (by PESTLE analysis)"

The analysed proposals were related to wind energy, desalination of seawater, biofuels production and collection of food oils. With this work it was possible to have a first implementation of the co-creation methodology previously trained, thus showing that it is feasible to apply in certain curricular units promoting the shift from classical education to a new approach centred on competencies of the reflective practitioner.

5 Conclusions

This paper presents a description of the application of the DEMOLA approach to IPS in Portugal. The DEMOLA® approach enabled the training of a group of IPS professors as facilitators of co-creation processes, having contributed to the acquisition of Design Thinking skills that can be used in new projects and applied to new teaching methods at IPS.

This approach also made it possible to strengthen IPS' partnerships with the national business networks which, in terms of internships and acceptance of IPS graduates in the labour market, creates a direct advantage.

The articulation between students and facilitators in the DEMOLA® approach contributed to a reinforcement of competencies at a national and international level, with companies and other polytechnic institutes, expanding the network of future collaborations.

To act as facilitators we identify the ability of self-management with learning and development based on personal demand, clearly evident during the 15 weeks of the DEMOLA programme.

Further studies cover the application of the co-creation design thinking process to other units from different curricular IPS programmes and develop a questionnaire survey conducted among students to better assess the impact of the used approach.

We hope to apply most of the learnings of this project in daily teaching practice, using active methodologies to respond to real problems.

The future is now!

References

1. F. Fernandez, Good practice case study - Demola. In *University Action Lab. Porto Business School* (2020)
2. EU. Open innovation, open science, open to the world a vision for Europe. European Commission (2016)
3. Demola Global, Demola Portugal Initiative (2021)

4. D. Catalá-Pérez, M. Rask, M. de-Miguel-Molina, The Demola model as a public policy tool boosting collaboration in innovation: A comparative study between Finland and Spain. *Technology in Society*, **63** (2020)
5. F.R. de Morais, J.B. Santos, Refinando os conceitos de cocriação e coprodução: resultados de uma crítica da literatura. *Revista Economia & Gestão*, **15**(40), 224–250 (2015).
6. V. Zwass, Co-creation: Toward a taxonomy and an integrated research perspective. *Intl. Journal of Electronic Commerce*, **15**(1), 11–48 (2010)
7. B. Leavy, Venkat Ramaswamy – a ten-year perspective on how the value co-creation revolution is transforming competition. *Strategy and Leadership*, **41**(6), 11–17 (2013)
8. C.K. Prahalad, V. Ramaswamy, Co-creation experiences: The next practice in value creation. *Journal of Interactive Marketing*, **18**(3), 5–14 (2004)
9. J. Trischler, S.J. Pervan, D.R. Scott, Exploring the "black box" of customer co-creation processes. *Journal of Services Marketing*, **31**(3), 265–280 (2017)
10. OCDE, Rural-Urban Partnerships (OECD Rural Policy Reviews) (2013)
11. J. Mazaj, S. di Bono, A.M. Li Destri, The role of local communities in the co-creation of innovations for inclusive and sustainable territorial development: the Madonie case in *10th International Scientific Conference Business and Management 2018* (2018)
12. L.F. Anderson, S.E. Robertson, Group facilitation: Functions and skills. *Small Group Behavior*, **16**(2), 139–156 (1985)
13. M. Luctkar-Flude, B. Wilson-Keates, J. Tyerman, M. Larocque, C.A. Brown, Comparing instructor-led versus student-led simulation facilitation methods for novice nursing students. *Clinical simulation in nursing*, **13**(6), 264–269 (2017)
14. C. Antuña-Rozado, J. García-Navarro, J. Mariño-Drews, Facilitation processes and skills supporting ecocity development. *Energies*, **11**(4) (2018)
15. S. Kaner, L. Lind, L.C. Toldi, S. Fisk, D.B. Foreword, M. Doyle, *Facilitator's guide to participatory decision-making* - Second edition (Sam Kaner, Ed.; Second Ed.). John Wiley & Sons, Inc. (2007)
16. L.Y.Y. Foong, M.B.M. Nor, A. Nolan, A. The influence of practicum supervisors' facilitation styles on student teachers' reflective thinking during collective reflection. *Reflective practice*, **19**(2), 225–242 (2018)

Influence of sales and value added on earnings in the transport and warehousing sectors

Jakub Horák[1,*] and *František* Jíra[1]

[1]Institute of Technology and Business in České Budějovice, School of Expertness and Valuation, Okruzni 517/10, 37001 Ceske Budejovice, Czech Republic

Abstract. Within the paper submitted, the author seeks the perspective for an investor, entrepreneur or anybody who does or plans to do business in the sector of transportation and warehousing and might need a method to determine the impact of sales and value added on earnings. The data are classified according to the CZ NACE coding for the transport and warehousing sectors. The complete data from the accounting records in the given field for the years 2016-2020 are selected. The mutual relationship of both parameters is determined using correlation. According to the results, in individual years, the correlation of both parameters and earnings differs, with each parameter showing a different impact on earnings in terms of individual years, type of company or region in which a given company operates. Overall, however, the research can be influenced by the given data, which need to be processed in more detail for possible further investigation of the given issue. The data were divided into parameters of individual years, regions, and the type of legal person.

Key words: Correlation, relevance, impact, field, data, transport, warehousing

1 Introduction

Transportation and warehousing play an irreplaceable role in the current world where goods and products are transported between different countries and even continents. There are many companies dealing with the issues of transportation and warehousing. Currently, these issues are referred to as logistics [1].

In recent years, there has been a very strong competitive pressure, with increasing customer demands for improved quality of services, speed of transport, and price. On the other hand, there are many factors that represent obstacles for logistic companies, including the receding Covid-19 pandemic, lack of qualified workforce, pressure to digitize and automate various processes, as well as the enormous increase in basic inputs. All these external factors have a significant impact on the earnings of companies. More than ever, these companies need to focus on their competitiveness, the ability to improve their performance and efficiency.

* Corresponding author: horak@mail.vstecb.cz

To answer these questions, senior managers need to be able to measure and compare results from various periods to track the achievement of business and economic goals. In evaluating the results, transportation and warehousing companies in general focus on using common financial analysis ratios, including profitability, debt, liquidity and the total assets, which express the earnings of a given company but primarily from a managerial and economic perspective. Accounting income and the derived indicators of classical financial performance may not be dependent on the creation of value for owners and investors. One of the main goals of companies in general is increasing the market value of the company over a longer period. This growth or decline can be measured using a number of methods where one of the most suitable ones is monitoring the so-called value added. Profitability analysis is an integral part of financial analysis of a company. Profit is one of the most commonly used indicators of business performance and is used to evaluate its historical performance. Due to their importance, it examines how the available information on company's profit can be used to increase profitability [2].

Given that especially in the transportation and warehousing sectors, the monitoring of value added could be neglected due to the monitoring of traditional accounting indicators, this paper focuses on analysing the impact of sales and value added on earnings.

This paper analyses the relationship between sales and earnings as well as the relationship between value added and earnings. These relationships are analysed for the current financial year in the transportation and warehousing sectors. The goal of the paper is thus to determine the appropriate profitability of the above ratios.

To achieve the goal of the paper, the following two research questions need to be answered:

- How do sales influence earnings in the transportation and warehousing sectors in the years 2016-2020?
- How does added value influence earnings in the transportation and warehousing sectors in the years 2016-2020?

2 Literature research

The evaluation of the success rate of companies in the selected sectors in the Czech Republic is closely related to the impact of the global financial and economic crisis which has also significantly affected the competitiveness of Czech companies [3]. Business success and value are complex concepts and the determination of the value of a business is conditioned by many internal and external factors that need to be considered in the evaluation [4]. One of the basic indicators is the earnings for a monitored period. Earnings are affected by a number of indicators that can be found in the financial statements, with the basic ones being sales and costs, which are further divided into other categories and sub-levels [5].

One of the fundamental sub-levels are sales. Sales refer to the basic information necessary for the calculation of profit or loss of a given accounting unit [6]. From an accounting perspective, sales are classified clearly. In some cases, problems may arise in the identification of sales and income from ordinary activities. This is one of the basic problems of financial accounting, as sales and income from ordinary activities determine earnings [7]. Sales determine revenues and represent one of the basic financial resources, which are then used to cover the expenditures and taxes of a given company. Especially in the industry, transport or trade sectors, sales are one of the main components of profit. Sales are determined by many internal and external factors, including e.g., the structure of the product mix in trading companies, production capacity in manufacturing companies, labour force capacity, or changes in the exchange rates in transport companies [8].

Earnings are also influenced by value added. There is a difference between so-called economic value added and book value added. One of the modern tools for evaluating business performance that has been recently paid great attention to is economic value added - EVA. The reported profit of any business has a very strong impact on the various business activities and decisions that the management of the company must make. The objective of earnings management is to report earnings in the amount required by the management of the company. Earnings management uses some creative accounting techniques [9]. EVA Equity is a difference between the return on equity and the opportunity costs of equity multiplied by the value of equity. This indicator is not only used to evaluate the company's ability to generate value for shareholders, but also to measure its success in the market, especially in global capital markets [10].

In contrast, book value added represents the difference between the value of a product and sales and the consumption within the production. In other words, the so-called book value added refers to the estimated difference between the production costs and output [11].

The amount of added value is a very important factor that predetermines the amount of net profit of a given company. The value added generated by a given company is used to cover wages, depreciation, repayment of loans and interest on these loans. The remaining amount then represents the net profit. Its amount is an important factor that influences profit or loss in earnings [12].

For manufacturing companies, it is important to maximize the use of production resources and optimize the production process in terms of production costs and time. Transportation plays an increasingly important role in the integrated logistics management of semi-finished products and component flows [13]. Sustainable transport management monitors two aspects of the issue – direct energy consumption of means of transport and indirect costs of resources. The issue of sustainable transport is considered a logical continuation of the sustainable development initiated by the UN General Assembly [14]. The problem may have risen in recent years, e.g., in public transport where transport authorities all over the world have reported about 95% decrease in the number of public transport users during the peak of the COVID-19 pandemic, along with a decrease in the revenues from fare boxes and additional costs of disinfection and implementation of the measures concerning physical distance [15].

Freight transport is showing a very clear shift towards eco-friendly means of transport [16]. In the long run, the development and modernization of various means of transport should not harm the environment [17], which has an impact on the costs of transport companies. There is still a characteristic dilemma between the economic efficiency of transport and transport services on the one side and political objectives on the other side [18].

Warehousing plays an important role in logistics and supply chain, with the operating costs accounting for more than 20 % of the overall costs of companies, which leads to the need for optimizing warehousing in order to reduce costs and increase profit. The objective of the transportation and logistics sectors is to reduce the costs associated with the distribution network. Often, there is also the need for determining the optimal ratio of warehouse location and connection to the transportation network [19].

It is thus necessary to look in more detail at the quality of sales and costs and analyse their mutual relationship to be able to predict future earnings [20].

A commonly used tool of financial analysis is ratio analysis, which provides a quick overview of the basic situation of a company in terms of its finances without complex and costly calculations. This analysis is based on mutual comparison and monitoring of individual components of the basic accounting statements. Commonly used financial statements include the balance sheet, profit and loss account, and/or cash flow statement. A large number of ratios can be mutually compared and compiled [21]. The economic situation of the company needs to be analysed in order to identify the need for change and to predict the future development of such a company. Financial analysis can be performed using a number of methods, with artificial neural networks being considered a very interesting and effective tool [22]. Financial ratios play an important role in determining corporate financial health, which helps to maintain the competitive position of an enterprise, with the achievement of stable development contributing to the elimination of potential financial risks [23].

Most studies are currently dealing with the ratios using the so-called Pearson correlation coefficient including sales and revenues. However, this can lead to neglecting some other indicators that generate positive earnings (profit) [24]. A study published in 2015 focused on analysing companies listed on the Stock Exchange in the period 2002-2011. The goal of the study was to examine the relationship between value added and accounting variables (net profit, operating profit) using Pearson correlation coefficient and the method of panel data regression.

The evaluation of business performance is an important aspect of nearly all economic decisions at the macroeconomic and microeconomic levels in the long and short term. The data concerning the attitude of partners towards the business and their interest in appreciation of investments can be interpreted using various indicators. It is important to understand the relationship between business performance and the amount of equity, while negative equity can be considered critical information threatening the existence of the company. The purpose of the quantitative research is to identify the relationship between the reported negative equity and business performance on a complete sample of financial data obtained from companies operating in the transportation and warehousing sectors in Slovakia with negative equity in the period 2014-2018. Business performance with negative equity is evaluated using Altman Z-score and the IN05 index, which enables classification of companies into zones – bankruptcy, prosperity, and grey zone. Pearson correlation analysis of the negative equity and the result of Altman Z-score confirm the existence of a strong direct relationship between negative equity and bankruptcy, weaker indirect relationship between negative equity and grey zone, and almost no relationship between negative equity and prosperity. As for the IN05 index, there is a low correlation between negative equity and all three zones. Companies with negative equity are in the zone of bankruptcy but do not have to necessarily cease their activities; they need to improve resource management, in particular to increase the amount of equity e.g., by generating profits from higher sales and good financial management [25].

Kaba and Svatosova [26] describe the procedures and methods for working with multivariate statistical datasets using the method of regression analysis. This analysis enables examining and describing the relationship and its form and express the identified dependence or dependencies by means of a mathematical function. This allows identifying a possible causal relationship between the variables and predicting the development of this relationship using the selected mathematical function when the acting of one variable enables determining the values of unknown variables [26]. In regression analysis, it is important to decide which regression functions to choose in order to identify the dependencies of the variables. The basic step is thus always the analysis of economic criteria and indicators [27].

For example, Svabova et al. [28] analyse the dependence of profit measures for the Visegrad Group using the methods of correlation analysis based on the data obtained from nearly 300,000 companies from the Amadeus database in the years 2013-2017. The authors performed a comparative analysis that identifies possible disproportions in the results identified for each of the analysed countries.

3 Data and methods

The data for the analysis will be obtained from the CRIBIS database of the company CRIF – Czech Credit Bureau. It is a database of companies operating in the transportation and processing sectors according to the classification CZ NACE. NACE is an abbreviation for the classification of economic activities established by the European Commission. It is the abbreviation of French Nomenclature statistique des activités économiques dans la Communauté européenn established in 1970 on the basis of international classification. For the purposes of this paper, section H – Transportation and warehousing according to the CZ NACE will be used, which includes freight and passenger transport, scheduled and non-scheduled transport, rail and road transport, transport by pipeline, water or air, and associated activities such as the activities of terminals, parking and warehousing facilities, trans-shipment points, etc. This section includes also renting means of transport with drivers or operators, as well as courier and postal activities.

The dataset includes complete information from financial statements between 2016 and 2020.

1) Companies in liquidation that are not so active will be excluded from the dataset.
2) There will be removed the columns that are not necessary for the calculations.
3) Value added will be calculated as follows:

$$\text{Trading margin} + \text{Outputs} - \text{Production consumption} \qquad (1)$$

Based on the profit and loss account, it is possible to determine:

a. Trading margin, which represents the difference between the selling and purchase price expressed in absolute numbers or as a percentage. The costs incurred for the sale of goods will be deducted from the sales of goods.
b. Outputs – this refers to the sum of sales of own goods and services, change in the status of in-house stocks of own production and capitalization
c. Production consumption – the sum of services and consumption of material and energy.
d. Data in the section "value added" with no values recorded are removed.

4) To find the answers to the research questions, it is necessary to determine the relationship between sales and earnings as well as the relationship between value added and earnings. The selected statistical method for determining these relationships is correlation analysis, which expresses the relationship between two variables using the correlation coefficient r. The given coefficient takes the values [-1;1]. The value of the absolute linear dependence is (+1) and (-1), while the value 0 expresses the absolute linear independence. If the coefficient has a value of -1, it is indirect dependence, while the value of +1 indicates direct dependence [29]. This method enables determining the dependence of earnings on sales and/or value added.

5) The results were further classified as follows:
   a) SALES
      i. By individual years
      ii. By regions within the CR
      iii. By the legal forms of individual companies
   b) VALUE ADDED
      i. By individual years
      ii. By regions within the CR
      iii. By the legal forms of individual companies

https://doi.org/10.1051/shsconf/202316001002

4 Results

4.1 Influence of value added on earnings

Table 1. The dependence by individual years.

Row labels	Correlation coefficient
2016	**0.80**
2017	**0.79**
2018	**0.81**
2019	**0.49**
2020	**0.61**

Source: Authors.

When analysing the individual years, it can be seen that the highest correlation between value added and earnings reaches the highest values in 2018, while the lowest values are reached in 2019. However, the correlation between value added and sales was confirmed.

Table 2. The influence by individual regions.

Row labels	Correlation coefficient
City of Prague	**0.66**
South Bohemian region	**0.42**
South Moravian region	**0.44**
Karlovarský region	**0.35**
Královéhradecký region	**0.62**
Liberecký region	**0.53**
Moravian-Silesian region	**0.61**
Olomoucký region	**0.54**
Pardubický region	**0.19**
Plzeňský region	**0.19**
Central Bohemian region	**0.55**
Ústecký region	**0.88**
Vysočina	**0.73**
Zlínský region	**0.72**

Source: Authors.

It follows from Table 2 that the lowest correlation is recorded in the Pardubický and Plzeňský regions, while the highest correlation is in the Ústecký region.

Table 3. The influence by legal form of companies.

Row labels	Correlation coefficient
Joint-stock company	**0.64**
Cooperative	**0.82**
European company	**0.84**
Spin-off company of a foreign legal person	**0.96**
Limited partnership	**0.00**
Limited liability company	**0.67**
State-owned enterprise	**0.86**
General partnership	**0.45**

Source: Authors.

The table of the influence by legal person shows the highest correlation between value added and earnings in the case of a spin-off company owned by a foreign legal person, while the lowest correlation is recorded in the case of general partnership.

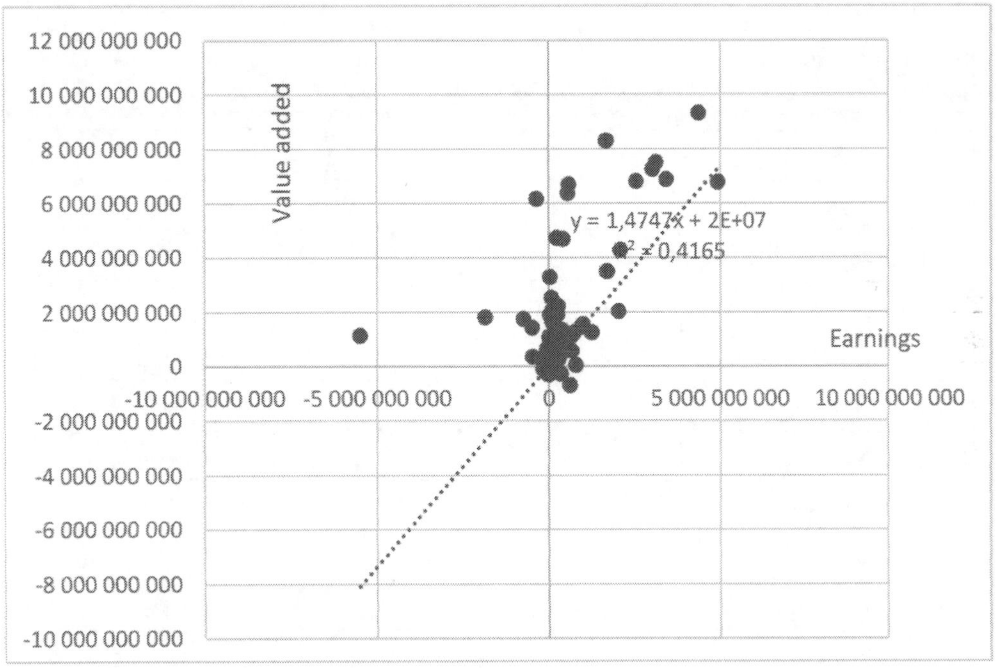

Graph 1. The influence of value added on earnings.

Source: Authors.

The resulting graph shows a linear trend, with the largest share of companies up to 2,000,000 in value added and 2,000,000 in terms of earnings.

4.2 Influence of sales on earnings

Table 4. The influence of sales by individual years.

Row labels	Correlation coefficient
2016	**0.36**
2017	**0.14**
2018	**0.33**
2019	**0.18**
2020	**0.15**

Source: Authors.

Sales and earnings show a relatively high volatility of correlation in individual years, and in principle, they do not correlate in any of the monitored years. The lowest correlation is recorded in the years 2017 and 2020. No direct influence on the specific market situation has been confirmed.

Table 5. The influence by individual regions.

Row labels	Correlation coefficient
City of Prague	**0.33**
South Bohemian region	**0.37**
South Moravian region	**0.40**
Karlovarský region	**0.26**
Královéhradecký region	**0.49**
Liberecký region	**0.46**
Moravian-Silesian region	**0.68**
Olomoucký region	**0.56**
Pardubický region	**0.88**
Plzeňský region	**0.29**
Central Bohemian region	**0.67**
Ústecký region	**0.64**
Vysočina	**0.89**
Zlínský region	**0.65**

Source: Authors.

In contrast, there is a higher correlation between sales and earnings in individual regions. The highest influence of sales on earnings can be seen in the Vysočina region, while the lowest one in the Karlovarský region.

Table 6. The influence by legal forms of companies.

Row labels	Correlation coefficient
Joint-stock company	**0.29**
Cooperative	**0.69**
European company	**0.85**
Spin-off company of a foreign legal person	**0.65**
Limited partnership	**0.05**
Limited liability company	**0.19**
State-owned enterprise	**0.55**

Source: Authors.

In terms of the influence of the legal form, there is almost no influence of sales on earnings in the case of limited liability companies and joint-stock companies. The highest influence of sales on earnings can be seen in the case of cooperative, European company, and spin-off companies owned by a foreign legal person.

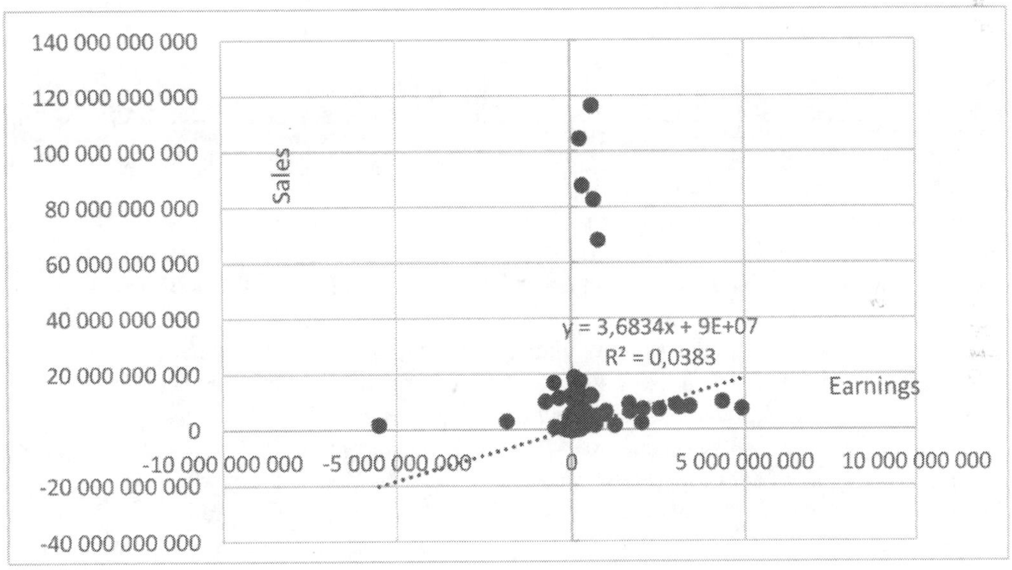

Graph 2. The influence of sales on earnings.

Source: Authors.

5 Discussion

The results of this paper were expected to provide answers to the formulated research questions:

- How do sales influence earnings in the transportation and warehousing sectors in the years 2016-2020?
- How does added value influence earnings in the transportation and warehousing sectors in the years 2016-2020?

The obtained and processed data show that in the transportation and warehousing sectors, a higher influence and correlation can be seen between value added and earnings rather than between sales and earnings. This trend is even more pronounced in the case of monitoring the influence by individual years.

In the context of monitoring the impact of sales by individual years, the biggest difference can be seen in the years 2017 and 2020 where almost no influence of sales on earnings is recorded. As for the year 2020, this could be due to the COVID-19 pandemic; however, given the situation in the year 2017 when the impact was even a tenth lower, this statement cannot be fully confirmed. On the contrary, data from publicly available sources and legislative changes do not show the reason for this increase compared to other years. Of course, this can be caused by some error in the data but the monitored sample was so broad that there would have to be several significant fluctuations in a given year, which is highly improbable. The most likely reason is that there was an error in the data; however, this cannot be verified in a completely relevant way.

Very interesting results were obtained in terms of analysing the impact of value added and sales on earnings by the type of company where limited liability companies and joint-stock companies were identified as the companies with the largest representation within CZ NACE classification. However, the highest correlation was identified in the case of spin-offs with a participation of a foreign legal person and European companies. In contrast, limited liability companies and joint-stock companies show the lowest correlation.

6 Conclusion

The aim of the paper was to analyse the relationship between sales and earnings as well as value added and earnings. These relationships were analysed in the transportation and warehousing sectors for a current financial year. The goal of the paper was thus to determine the optimal profitability for given indicators.

It can thus be stated that the goal of the paper was achieved. The benefits of this paper include the information for investors in the transportation and warehousing sectors on which type of companies and in which regions sales and value added can influence earnings most. From the perspective of individual years, this view is not completely relevant, because it considers past years and the data are available only up to the year 2020 when the COVID-19 pandemic started, which might have significantly influenced the results. This cannot be fully confirmed as globally, there is no historical experience of a similar situation.

Nevertheless, the results obtained can be used as a background for further research in the given sector. There is a lot of space for further research that could analyse the obtained data in more detail, specifically in terms of the fluctuations observed in individual years. A limitation of the research could be the analysed data, which may have not been correctly recorded during their collection. Also, the crucial year 2021 was not included, which is the year of the experience with the COVID-19 pandemic. This could have significantly affected the results obtained.

Acknowledgements

This research was supported by the project, which has received funding from the Institute of Technology and Business. Grant Number IVSUZO001, "Benchmark of the world economy in the context of the principles of the circular economy and determining the value of the company". This paper was also supported from the project of Specific University Research, reg. number 05SVV22 – Scientific School and its connection with the experts from practice in the research activities of the institution.

References

1. P. Krawczyk, P. Kokot-Stępień, The impact of the exchange rate on the financial result of enterprises in the transport sector. *Ekonomia i Prawo,* **19**(1), 47-60 (2020)
2. E. Kalinova, P. Mrazova, Profitability of company in industry. *Ad Alta-Journal of Interdisciplinary Research,* **12**(1), 216-219 (2022)
3. M. Pech, J. Prazakova, L. Pechova, The evaluation of the success rate of corporate failure prediction in a five-year period. *Journal of Competitiveness,* **12**(1), 108-124 (2020)
4. E. Adamikova, T. Corejova, Creative accounting and the possibility of its detection in the evaluation of the company by expert. *Journal of Risk and Financial Management,* **14**(7), 327 (2021)
5. S. Bojnec, S. Zampa, Subsidies and economic and financial performance of enterprises. *Journal of Risk and Financial Management*, **14**(11), 1-16 (2021)
6. A. Vighova, Accounting aspects of sales in double entry accounting of Slovakia and comparison with international accounting standards. *Entrepreneurship and Sustainability Issues,* **9**(2), 81-93 (2021)
7. H. Buk, Trends of Changes in recognition of companies' sales from ordinary activities. *Problemy Zarzadzania,* **16**(72), 176-184 (2018)
8. M. Vochozka, J. Vrbka, V. Stehel, P. Suler, Z. Rowland, V. Machova, J. Horak, T. Krulicky, *Finance podniku: komplexní pojetí [Corporate finance: a comprehensive concept].* Prague: Grada Publishing (2021)
9. L. V. Popova, I. A. Maslova, Y. A. Kotlova, M. V. Selyukov, Z. V. Mkrtchyan, Accounting and analytical system of value added from the position of controlling. *Imitation Market Modelling in Digital Economy: Game Theoretic Approaches,* **368,** 702-710 (2022)
10. V. Machova, P. Suler, The influence of value added and economical value added in the construction industry. *17th International Scientific Conference on Globalization and Its Socio-Economic Consequences,* 1436-1443 (2017)
11. J. Jilek, *Nástin sociálněhospodářské statistiky [Outline of socio-economic statistics].* 2. ed. Prague: Oeconomica (2005)
12. I. Neumaierova, I. Neumaier, *Výkonnost a tržní hodnota firmy [Performance and market value of the firm].* Prague: Grada. Finance (2002)
13. J. Krystek, S. Alszer, Analysis and simulation of internal transport in the high storage warehouse. *Analysis and Simulation of Electrical and Computer Systems,* 407-420 (2018)
14. O. Shirokova, L. Shirokov, Management and economic performance of sustainable transport. *International Scientific Conference Energy Management of Municipal Transportation Facilities and Transport EMMFT,* 310-317 (2018)
15. R. Naveen, A. Gurtoo, Public transport strategy and epidemic prevention framework in the Context of Covid-19. *Transport Policy,* **116**, 165-174 (2022)
16. C. Tawfik, S. Limbourg, Pricing Problems in intermodal freight transport: Research overview and prospects. *Sustainability,* **10**(9), 3341 (2018)
17. S. E. Rudneva, Future development of transport and environment. *IOP Conference Series: Materials Science and Engineering,* **941**(1) (2020)
18. D. Ballarano, S. M. Patella, F. Asdrubali, Sustainable transportation for events: A systematic review. *Sustainability*, **14** (2022)
19. P. V. Popov, I. Y. Miretskij, Methodology for constructing the region's logistics infrastructure. *Economy of Region*, **15**(2), 483-492 (2019)
20. P. Bilinski, M. Eames, Analyst revenue forecast reporting and the quality of sales and expenses. *Journal of Business Finance & Accounting*, **46**(1-2), 136-158 (2019)

21. A. Knapkova, D. Pavelkova, D. Remes, K. Steker, *Finanční analýza: komplexní průvodce s příklady [Financial analysis: a comprehensive guide with examples].* 3. ed. Prague: Grada Publishing (2017)
22. T. Krulicky, Using Kohonen networks in the analysis of transport companies in the Czech Republic. *Innovative Economic Symposium 2018 - Milestones and Trends of World Economy (IES2018),* **61** (2019)
23. T. Kliestik, K. Valaskova, G. Lazaroiu, M. Kovacova, J. Vrbka, Remaining financially healthy and competitive: The role of financial predictors. *Journal of Competitiveness*, **12**(1), 74-92 (2020)
24. Q. Zhou, H. Liu, W. Li, T. Mo, B. Wu, Bilateral autotrading framework for stock prediction. *International Joint Conference on Neural Networks (IJCNN),* 1-8 (2021)
25. K. Tasaryova, R. Paksiova, The impact of equity information as an important factor in assessing business performance. *Information*, **12**(2), 85 (2021)
26. B. Kaba, L. Svatosova, *Statistika [Statistics].* 3. ed. Prague: Czech University of Agriculture (2001)
27. R. Hindls, S. Hronova, J. Seger, *Statistika pro ekonomy [Statistics for economists].* 4. ed. Prague: Professional Publishing (2003)
28. L. Svabova, K. Valaskova, P. Durana, T. Kliestik, Dependency analysis between various profit measures and corporate total assets for Visegrad Group's business entities. *Organizacija,* **53**(1), 80-90 (2020)
29. M. Synek, H. Kopkane, M. Kubalkova, *Manažerské výpočty a ekonomická analýza [Managerial calculations and economic analysis].* 1. ed. Prague: C.H.Beck (2009)

Changing patterns of business behavior: The Ukrainian experience of adaptation of global accounting and financial mechanisms

Tetyana Calinescu[1], *Ganna* Likhonosova[1,*], and *Olena* Zelenko[2]

[1]National Aerospace University "Kharkiv Aviation Institute", Finance, accounting and taxation Department, 61070, 17 Chkalova St., Kharkiv, Ukraine

[2]Volodymyr Dahl East Ukrainian National University, International Economics and Tourism Department, 49010, 74 Gagarin Ave, Kyiv, Ukraine

Abstract. In the context of growing turbulence in economic development, it becomes necessary to determine the main patterns of interaction between business and its counterparties. The purpose of this study is to determine behavior patterns - patterns and eliminate existing obstacles to leveling negative phenomena in the economy and finding conditions for meeting business needs, incentives for further development, improving the quality and standard of living, attracting more financial resources, investments to ensure production. The methodological basis of the study was modern regulatory and legislative sources of formation of accounting and financial mechanisms and patterns in business activity. In the conducted researches, methods of dialectics, methods and principles of scientific knowledge. As a result of the study, the possibility of substantiating the patterns of business behavior that form the system of financial, transactional management and strategic accounting of intellectual property and its structural components based on a complex of conceptual approaches is determined: the theory of intellectual capital, intellectual property valuations, accounting and management, the methodology of net intellectual capital accounting. Conclusions are presented by the optimal variant of pattern changes that corresponds to the selected conditions of adaptation of the accounting and financial mechanism to the conditions of business development can be obtained only on the basis of intuitive selection, appropriate traditional methods, algorithms, strategies, which will be the basis of the set goals and tasks related to the nature of doing business.

Keywords: pattern, trend, template, system, finance, changes, development

1 Introduction

According to research by international institutions, the development of technologies in the world is exponential, which means that every year the number of changes that occur per unit of time is constantly increasing. Therefore, in order to survive, business must have time to adapt, experiment with new approaches, world models and accounting and financial

*Corresponding author: a.likhonosova@gmail.com

mechanisms [1]. At the same time, 70% of the risks in a change situation are not related to technologies and processes, but to employee sabotage, overt or hidden. Therefore, the success of all changes in business depends on changing the patterns and models of human behavior (smokers, producers and consumers), which is subject to the general global socio-economic and accounting and financial mechanisms and trends, the main of which is the increase in internal and external contradictions of the modern way of social production, gaps between the technical, economic and humanitarian spheres of development, which are projected and perceived directly by man. That is, a person is influenced by the dominant trends that exist in his environment and affect those patterns that were formed during certain periods and circumstances. However, the surrounding world and man are constantly developing, accumulating experience, self-improvement. Patterns are transformed accordingly, because modern trends towards harmonious business development, mastering new, more effective skills force new shifts, changes in existing development mechanisms to more active ones or replacement of some non-working tools with more productive ones.

The fact is that not all methods of change and movement in society can be comfortable for business, because in the process of adapting some accounting and financial mechanisms, especially those that worked in some countries, but do not work in Ukraine (for example, financial "shock therapy ", which worked in Poland, but not in Ukraine or the pension reform in Chile, which still does not work in Ukraine), can bring some unpleasant consequences. Therefore, there is a need to analyze patterns that will help to better understand business behavior and allow to more usefully and efficiently adapt new or modernized accounting and financial mechanisms to existing situations, facts and adjust them to conscious business behavior.

In the last 20 years, due to the rapid development of technologies, the role of the relationship between the client and the company has increased many times. Consumers increasingly need the trust that comes from a successful experience with a seller. However, by shifting the focus from the product to the customer, investing in a unique interaction experience and trying to satisfy all customers at once, the company risks becoming unprofitable. Therefore, a strategic approach must be used to successfully implement a balanced client-centric business model. Businesses should allocate investment in customer relationships based on the value they bring. Most often, it is characterized by current and potential indicators of profitability and turnover, but there may be other criteria depending on the company's strategy.

Behavioral patterns in the business community are most often considered as "patterns" that customers follow when making decisions about buying or interacting with a company. Some of these patterns lead to mutually beneficial cooperation and high profitability, while others lead to customer churn or long unprofitable relationships. The task of a business unit is to continuously incentivize the "right" behavior, or in other words, to initiate behavioral patterns that affect its profitability and strategic goals. Although, we should not forget about those general trends that form "habits" in the behavior of business agents themselves and contribute to the improvement of their financial situation.

Based on this, the purpose of this study is aimed at eliminating existing obstacles to eliminating negative phenomena in the economy and finding conditions for meeting business needs, incentives for further development, improving the quality and standard of living, attracting more financial resources, deposits and investments to ensure production. This goal acquires special attention and urgency in the conditions of active military actions taking place in Ukraine today and the processes of economic reconstruction in the territories that were cut off from the production complex of Ukraine.

In this context, the use of foreign experience [2, 3, 4, 5, 6, 7, 8, 9] and the application of the existing practice of accounting and financial mechanisms, legislation and EU standards are the basis for rethinking the state budget and financial policy in Ukraine, in the context of ensuring and protecting working businesses in order to form the most effective regulatory mechanisms and growth of the national economy.

2 Methodology

The methodological basis of the study was modern regulatory and legislative sources of formation of accounting and financial mechanisms and patterns in business activity. In the conducted researches, methods of dialectics, methods and principles of scientific knowledge, tools of economic and comparative analysis were used for the successful adaptation of foreign experience in the activity of the Ukrainian business environment.

With the help of comparative and statistical methods, expert assessments of losses and losses in the business sector as a result of the transformation of business behavior models were summarized, trends in the directions of industry restructuring and the level of business activity of Ukrainian enterprises were analyzed. Historical and logical approaches are used to generalize and systematize the possibilities of using in Ukraine the experience of countries that were participants in military conflicts, regarding the ways of business recovery and ensuring the functioning of accounting and financial mechanisms.

The study of management tools of the accounting system of Ukrainian enterprises determines the need to use the fundamental principles of world economic science: regional-spatial economics (from the standpoint of implementation opportunities in different territories) and management of economic entities, which in aggregate involves the use of theoretical and methodological approaches, special and scientific methods. In particular, the methods of scientific abstraction and synthesis are used to identify and group the main external and internal threats to transaction management and strategic accounting, to systematize opportunities for development and to ensure the security of Ukrainian enterprises; the modeling method is used to substantiate the logic of conducting business investigations regarding the use of borrowed experience in the application of accounting and financial mechanisms; monographic analysis made it possible to determine the essence of the process of changing business behavior models with the possibilities of synergistic cooperation of regional clusters; a systematic approach is used for a comprehensive assessment of socio-economic transformations in Ukrainian business and their consequences; reflexive methods are used to compare management algorithms for compliance with levels of business behavior to modern conditions; special methods of historical and logical analysis used to determine the main stages of the Data mining algorithm within the framework of the management system.

However, Lois P., Drogalas G., Karagiorgos A. and Karasteriou E. note the emergence of the phenomenon of unpredictable financial behaviour, which is expressed through tax evasion due to undeclared employment [4].

An unusual analogy for the impact of greening processes on financial stability in the world is given by the authors Jadoon I., Mumtaz R., Sheikh J., Ayub U. and Tahir M., emphasizing the importance of investing in green energy on the financial performance of multinational corporations [10].

Martini A. conducts a detailed study of the importance of global socially responsible investment from the standpoint of the dynamics of financial behaviour of business units. Such financial activity has intensified in recent decades, from a religious niche in the world to a core risk analysis strategy for institutional and retail investors. Research shows that the lack of a globally accepted taxonomy of what is a sustainable financial activity coupled with practical and behavioural difficulties are major critical issues that deter the socially responsible investment industry globally [9]. Similar aspects, but in terms of human capital attributes, are studied by Lamboglia R. and Mancini D., emphasizing the achievements and possible shortcomings of regulatory activity in this area at the international level in recent years with special emphasis on the European Union [8].

This study takes place within the framework of scientific research (research and design) work, which is carried out on the basis of the Department of Finance, Accounting and Taxation, National Aerospace University "Kharkiv Aviation Institute" on the topic "Accounting and financial mechanism of innovative interaction of business entities in the context of the transformation of the knowledge economy and socio-behavioral restrictions" (state registration number 0121U108324).

3 Results

Already at the beginning of 2021, marketing experts identified the main trends in the behavior of representatives of the business environment, which indirectly affect the organization, stability and success of financial and accounting processes, most of them have not lost their relevance even now [11]:

1. Forming the habit of being attentive always and in everything will become a necessity in everyday work. The vitality and success of any enterprise is subject to a lot of different factors, but the most successful will be those who can predict the impact of these factors and prevent the impact of negative ones, enhancing the positive impact. Actions to anticipate problems and prepare for crisis phenomena, even if officially available information does not confirm the possibility of their appearance, is the first pattern that is important in the current year and in the future, which will allow maintaining financial stability in the long term.

2. Downsizing and downsizing is a trend that is becoming a pattern for most representatives of the business environment: reducing the number of events, reducing the number of participants, minimizing and simplifying content on websites, shorter videos, smaller volumes of printed materials, smaller sizes of things. Transition from large-scale events to small meetings in a narrow circle; giving preference to simple, small, understandable and convenient things, a simplified way of life - the need formed in 2020 gradually turns into a pattern for both the business itself and society as a whole, creating the basis for avoiding excessive costs and saving financial resources.

3. Practical application of artificial intelligence - in the last two years, the use of artificial intelligence in everyday life has been actively spreading both among the population and in the business environment. It greatly simplifies everyday work with typing. Developed software products facilitate the creative process of creating texts, assisting in editing and generating them independently based on predefined parameters. In order to speed up work processes, most entrepreneurs should involve such a tool in the organization of their activities and actively use it - this is the third general pattern for business entities.

4. Investing in human capital is another modern trend. Qualified specialists are always a valuable resource for any enterprise, but now we are talking about investments in existing employees, this is the pattern that has been forming for the past two years and is only now being consolidated: maintaining and developing the professional competencies of permanent employees is much more profitable from a financial and economic point of view vision (and not only) than the search and attraction of highly paid specialists.

5. Individualization of video materials. Video as a form of content will continue to gain momentum, but its content is becoming more differentiated. This also applies to video conferences: from mass video communication programs, professional users move to individual communication, including the creation of individual video material for each partner, client, counterparty, etc., because it is easier to watch and listen than to read. Such training actually turns into a new way of communication that can save time and, accordingly, money.

6. Adaptation of web content to the needs of mobile users. Active promotion of your own business and financial and economic stability will be impossible if the company does not adapt its site for mobile devices. The majority of potential counterparties, customers are no longer ready to spend time opening a computer, it is much easier to pick up a phone and perform an operation in the mobile version of the site, through the appropriate application, etc.

7. Greening of businesses. At first glance, this trend has nothing to do with the improvement of the financial and economic stability of subjects of economic activity, but this is only at first glance. The modern civilized world is aware of the scale of environmental problems created by mankind during its existence. Greening becomes the main need and condition for the survival of the planet. The necessity of this trend was confirmed by the recorded facts of the regeneration of ecosystems due to the cessation of production, a reduction in the number of visitors to famous tourist destinations, etc. during quarantine measures in various parts of our planet. The direct interdependence between the environmental friendliness of production and business processes and the commitment of partners, customers, and counterparties becomes obvious.

The presented world trends of behavior in the economic environment are necessarily, one way or another, reflected in the financial position of any enterprise, taking into account these trends is an important component of the formation of a positive financial flow.

Segmentation will help determine business behavioral patterns in the new Ukrainian realities. As a starting point, you can take the selection criteria that are already used at the enterprise - based on consumer needs, socio-demographic characteristics, etc. - and then add additional parameters and accounting and financial mechanisms for their processing. These include those that reflect the characteristics of customer behavior: frequency of purchases, rate of repeat purchases, average check, best accounting and financial channels.

In addition, it is necessary to take into account the cost of servicing the segment and the presence of requirements for customization of the product offered to the consumer. Having collected such an array of data, it is possible to form segments based on customer behavior, the basis of which is strategic segmentation. Each segment should be evaluated, at least, according to the following characteristics: profitability, speed and convenience of accounting and financial mechanisms, number of customers and average duration of relationships.

Conducting such an analysis allows you to identify behavioral patterns that have the greatest impact on profitability. In this way, the company will understand exactly how the consumer can behave and what actions he can take to increase his value for the business. Examples of behavioral patterns that positively affect the development of accounting and financial mechanisms for ensuring the profitability of the enterprise:

- making repeat purchases;
- purchase of an annual subscription;
- periodic repeated identical purchases;
- assistance in generating new customers;
- more frequent use of the online communication channel.

Behavioral patterns that negatively affect the development of accounting and financial mechanisms for ensuring the profitability of the enterprise include:

- interaction with the enterprise only through offline representative offices;
- purchase of low-margin products;
- many requests for technical support;
- frequent use of promotional codes;
- product return.

One or more metrics should be assigned to each such pattern and their current and potential value should be calculated. Not all metrics will be direct, sometimes they may be indirect. Thanks to this, it will be possible to control how effectively CRM programs influence the change in customer behavior. Also, each pattern needs to be assigned one of the basic CRM metrics: profitability, speed and convenience of accounting and financial mechanisms, number of customers and average duration of relationships.

In order to develop new effective patterns of business behavior in accordance with the selected experience of applying accounting and financial mechanisms, it is possible to suggest following the following recommendations [14]:

1. Each experience or situation should be considered from different angles, analyzing not only the pros, but also the cons in order to make the right decision or create a new position;

2. Any business environment is multifaceted, polarized, so a characteristic feature for many types of business will be that they will not be able to fit their behavior into established and already existing forms and frameworks of patterns;

3. In order to solve the existing problems, a moderate choice should be made regarding whether the business is able to independently get out of the state of helplessness through the development of creative entrepreneurship, or whether it should seek help from other business entities.

4. Each pattern of business behavior can use several fundamentally different approaches: destroy the old behavior, add a more successful new one, or simultaneously destroy the old and create a new one. However, it is considered better to add new ones on the basis that the business (such as the person running the business) will use and test the new offerings. But for each moment of time and situation, there is only one best option for the unfolding of events, which can be found in the process of a certain conscious selection, because making the right choice is a state of formed consciousness and a business attitude to the norms and rules of behavior that exist in society.

5. In its daily activities, business is guided by certain patterns of behavior that are learned at the unconscious level. However, it is important to focus on patterns related to the area of development interest.

6. When working with patterns, it is important to understand that it is extremely difficult to change them without the help of specialists, especially if we are talking about accounting and financial mechanisms and the need to adjust them to the military conditions of the country's economic development, and it requires the business to have at least basic knowledge.

According to A. Allen, an employee of the American Institute of Labor, the accounting of business behavior is: the development of theoretical foundations that explain the nature and determine the factors of the value of people from the point of view of official organizations; development of reasonable and reliable methods for assessing the value and value of people for organizations; design of organizational support for the implementation of the proposed assessment methods [12] based on the use of indicators of net assets and net liabilities in adequate assessments.

Methods of measuring and evaluating business behavior are focused on the use of accounting engineering tools [13]. The pattern of business behavior and its accounting methodology is focused on the use of an integrated system of financial, managerial and strategic accounting with transactional managerial and strategic accounting (Figure 1).

The system of integrated accounting with transactional management and strategic accounting is built on a structured plan of accounts with integration on three levels [13]:

1. Financial accounting, based on the financial accounting interface, forms the basis of the system and the use of a structured chart of accounts is being built;

2. A system of transactional financial accounting, which includes computer programs embedded in a structured chart of accounts: 1) a structural derivative balance sheet; 2) marketing derivative balance sheet; 3) behavioral derivative balance sheet; 4) segmental derivative balance sheet; 5) forecast derivative balance sheet; 6) strategic derivative balance sheet; 7) intelligent derivative balance sheet. At the same time, sections of the chart of accounts, mega accounts, net assets as actual or forecast data can act as the initial operator.

3. Resource potential management system of the enterprise: hedged; integrated risk; monitoring derivative balance sheets and others.

The transactional accounting system is a modern version of creative accounting that provides management of the enterprise's resource potential. The issue of creative accounting, which has come to the fore again, is the modern equivalent of the previously existing distortions of reporting information - profit smoothing, selective reporting of data, as well as pressure from companies to prevent the introduction of standards in certain areas.

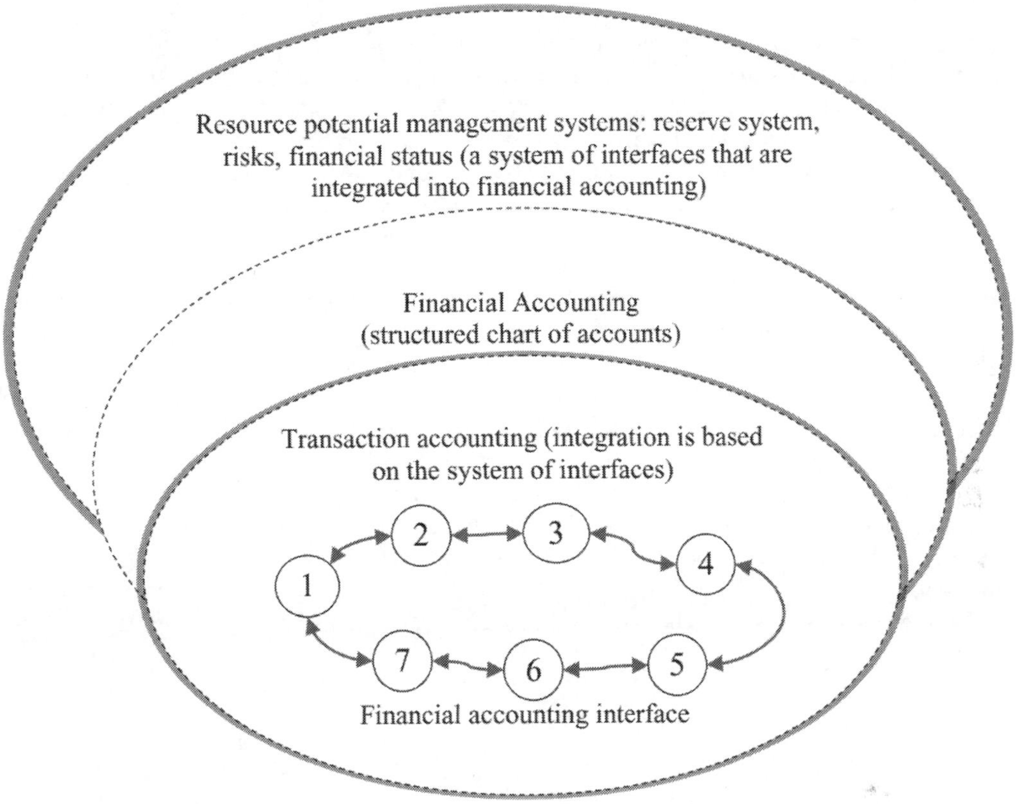

Fig. 1. Integrated accounting system with transactional management and strategic accounting. Source: developed by the authors based on [10].

But not always existing or foreign accounting and financial mechanisms, as the practice of the author's studies [15, 16, 17, 18] have shown, work in the conditions of Ukraine. For the modern period, the most relevant mechanisms are those that transform the debt model of deficit financing of the economy into the necessary attraction of the necessary financial resources to ensure production, solve the problem of quality of life and economic growth [19]. However, mechanisms due to the motivation to invest funds in the development of production, financial operations that ensure higher profitability have not lost their relevance. And in the context of the European integration of Ukraine, EU standards and practices are applied to ensure social rights, receive fair wages that ensure a decent standard of living and prevent poverty. Rethinking the state policy regarding incomes in the context of provision, protection of the working person, devaluation of labor, will allow to form the most healthy prerequisites for the economic growth of business and activate the creation of new accounting and financial mechanisms. In addition, the experience of integration processes shows that for a long time there has been a gradual harmonization of the accounting systems of EU member states by adapting national legislation and harmonizing accounting and finance methodologies to the conditions of Ukrainian business development. The experience of harmonization of accounting and financial mechanisms for

EU countries determines the list of priority tasks, which is not only an important prerequisite for strengthening economic ties between Ukraine and the EU in the near future, but also conditions the formation of new patterns

4 Discussion

Any business is done by proactive people who achieve results, know how to think strategically and create patterns of behavior regarding accounting and financial mechanisms that are adapted together with representatives of various businesses. Although we do not have to copy and adjust our patterns to those that already exist in the world, but we have to apply and shape our approaches [20], but if we examine the existing patterns of business behavior and take them as a basis, then those patterns attract attention thanks to which success has already been achieved or these achievements can be repeated. Therefore, based on the conditions of development in which Ukraine is today (conduct of active military operations and further reconstruction of the country's economy), it is sufficient to choose those models of behavior that already have a certain image (authority) and apply them to identical, similar circumstances that have developed in business With the development of global and domestic financial systems, the number of patterns for adapting certain accounting and financial mechanisms is increasing. Every business has open access to and to companies highlighting their positive developments in this area. However, it is important to remember the need to conduct a certain investigation regarding the use of information and the reliability of the provided data in order to confirm the probability of avoiding risks through the application of the proposed tools to the conditions of business activity, to assess the scope of the use of borrowed experience and to predict possible achievements (effects, including synergistic ones). The sequence of such investigations is presented in fig. 2.

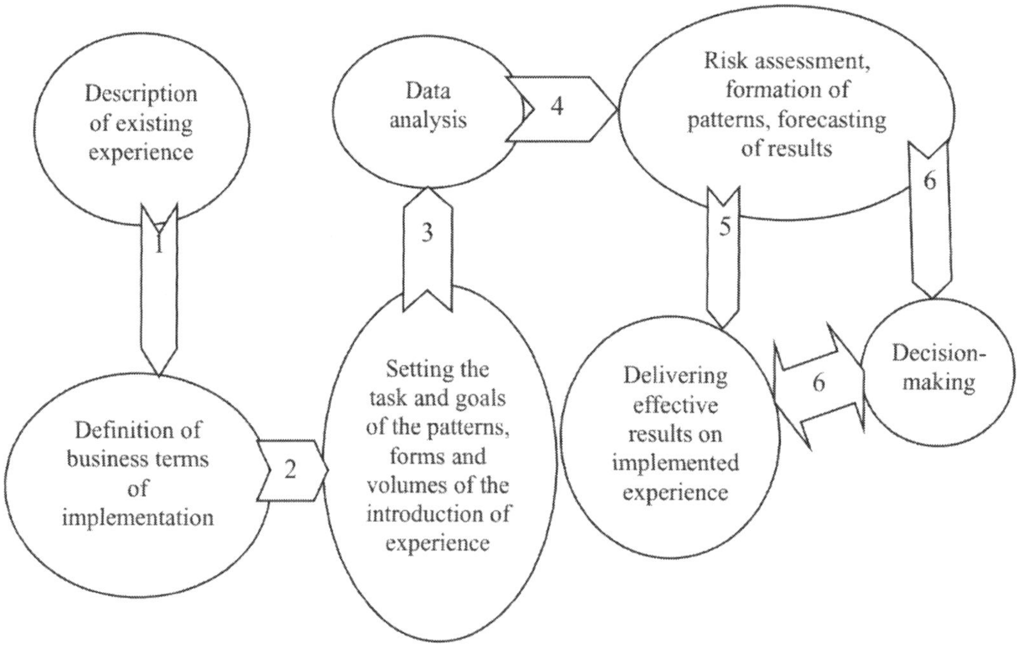

Fig. 2. The logic of conducting business investigations regarding the use of borrowed experience in the application of accounting and financial mechanisms.
Source: developed by the authors using [21].

When forming new patterns (or their transformation) regarding the effective use or adaptation of accounting and financial mechanisms, in addition to ascertaining the facts according to the logic of conducting business investigations (Fig. 2), it is necessary to predict business development trends and the emergence of all possible threats, advantages and disadvantages, to determine diagnosis, conduct data mining or SWOT analysis to obtain the information presented in the Table 1.

Table 1. Main stages (algorithm) of using Data mining or SWOT analysis within the framework of the system.

Analysis of the current situation (diagnosis)	Analysis of future actions
1. Analysis of critical internal and external factors	
2. Evaluation of external factors (expertise)	5. Forecasting development trends for each determined external factor
3. Assessment of internal factors (expertise)	6. Development of recommendations to strengthen the advantages and eliminate the disadvantages of each of the internal factors
4. Determination of diagnosis: specific positions and state of business, disadvantages and advantages)	7. Forecasting the future (prospective vision)
	8. Formulation of patterns and directions of their change based on each direction of business activity

Source: developed by the authors based on [22].

Stages, which are presented in the Table 1 provide clear recommendations on the order of analysis within a specific system and allow you to identify all problem areas that need to be paid close attention to.

5 Conclusion

The conducted research determined the possibility of substantiating patterns of business behavior that form a system of financial, transactional management and strategic accounting of intellectual property and its structural components on the basis of a complex of conceptual approaches: the theory of intellectual capital, valuations of intellectual property, accounting and management, the methodology of accounting for net intellectual capital; transactional accounting methodology (accounting and methodological support, structured chart of accounts, accounting architecture, information transactions, integrated accounting system); management systems of the reserve system of the enterprise, intellectual capital, risks, financial situation.

As the research has shown, the optimal variant of pattern changes that meets the selected conditions of adaptation of the accounting and financial mechanism to the conditions of business development can be obtained only on the basis of intuitive selection, appropriate traditional methods, algorithms, strategies, which will be the basis of the set goals and objectives, related to the nature of business. Undoubtedly, all existing and proposed methods of analysis, evaluation and other proposed methods of making the right decisions based on Data mining or SWOT analysis can be envisaged for the expected development of the economy of Ukraine. However, the economy of Ukraine today is developing in unpredictable ways, which creates certain additional problems regarding changes in business behavior patterns. And there is hope that the application of the proposed decision-making tools, the transformation of patterns and the adaptation of already existing experience in the application of accounting and financial mechanisms will create an opportunity for the emergence of new approaches and innovations in the functioning of not only business in the conditions of military operations, but also the entire Ukrainian society, which is united for fulfillment of the tasks set for the liberation of the occupied territories, business development, and the transition from wartime conditions to peaceful production, which involves integration into the European and world economic community, the financial system, and acceptance of the acquisition of the world economy and civilization as a basis.

References

1. Change to live in the future. 5 Life Hacks – How to manage changes in the company [online], Available at: https://mim.kyiv.ua/publication/zmnyuy-schob-jiti-v-maybutnomu (2019)
2. R. Nanfosso, J. Hadjitchoneva, The European Union facing the challenges Globalisation. *Economic Studies,* **30**(5), 27 – 47 (2021)
3. S. Barrios, D. d'Andria, M. Gesualdo, Reducing tax compliance costs through corporate tax base harmonization in the European Union. *Journal of International Accounting, Auditing and Taxation*, **41** (2020)
4. P. Lois, G. Drogalas, A. Karagiorgos, A. Thrassou, D. Vrontis, Internal auditing and cyber security: audit role and procedural contribution. *International Journal of Managerial and Financial Accounting*, **13**(1), 25 – 47 (2021)
5. M. Mandel, J. Vejmalěk, Testing the properties of financial analysts' predictions of future spot exchange rates (Example of CZK/EUR). *Finance a úvěr-Czech Journal of Economics and Finance,* **71**(1), 33 – 51 (2021)
6. N. Albu, C. Albu, S. Gray, Institutional factors and the impact of international financial reporting standards: the Central and Eastern European experience. *Accounting Forum,* **44**(3), 184 – 214 (2020)
7. A. M. Nasir, R. Zainudin, Aidil R. Shahrin. Financial revenues, financial leverage, debt maturity, uUncertainty and the underinvestment problem. *Finance a úvěr-Czech Journal of Economics and Finance*, **72**(1), 71 – 101 (2022)
8. R. Lamboglia, D. Mancini, The relationship between auditors' human capital attributes and the assessment of the control environment. *Journal of Management and Governance*, **25**(4), 1211 – 1239 (2021)
9. A. Martini, Socially responsible investing: from the ethical origins to the sustainable development framework of the European Union. *Environment, Development and Sustainability*, **23**(11), 16874 – 16890 (2021)
10. I. Jadoon, R. Mumtaz, J. Sheikh, U. Ayub, M. Tahir, The impact of green growth on financial stability. *Journal of Financial Regulation and Compliance*, **29**(5), 533 – 560 (2021)
11. J. Jantsch. 7 Small business trends that arrived just in time for 2021. Duct Tape Marketing [online], Available at: https://ducttapemarketing.com/small-business-trends-2021/ (2021)

12. A. Allen, M. F. Lewis-Western, K. Valentine, The innovation and reporting consequences of financial regulation for young life-cycle firms. *Journal of Accounting Research*, **1**(60), 45-95 (2021)
13. V. Druzhynina, G. Likhonosova, G. Lutsenko, Assessment welfare of the population in the synergetic system of socio-economic exclusion. *Marketing and Management of Innovations*, **2**, 54–68 (2018)
14. Pattern – what is it in psychology? Behavior patterns [online], Available at: https://www.kozaky.org.ua/patern-shho-ce-take-v-psixologi%D1%97-paterni-povedinki/ (2020)
15. T. Calinescu, G. Likhonosova, O. Zelenko, International financial activities: accounting, taxation and insurance. *Baltic Journal of Economic Studies*, **8**(2), 83 – 90, (2022)
16. T. Calinescu, G. Likhonosova, O. Zelenko, Estimation of conditions the realization of democratic mechanisms the transformation of society: tearing away and social dialog. *Montenegrin Journal of Economics*, **14**(1), 93-107 (2018)
17. T. Calinescu, G. Likhonosova, O. Zelenko, Parameters of the region's economic development of Ukraine in the context of global financial change. Innovation management and information technology impact on global economy in the era of pandemic. *Proceedings of the 37th International Business Information Management Association Conference (IBIMA) 30-31 May 2021*, Cordoba, Spain. By ed. Khalid S. Soliman, **7**(16), 4803 – 4812 (2021)
18. T. Calinescu, G. Likhonosova, O. Zelenko, Regulators of financial invulnerability of Ukraine: International factors of influence and accounting methodology. Sustainable economic development and advancing education excellence in the era of global pandemic. *Proceedings of the 36th International Business Information Management Association Conference (IBIMA), 4 – 5 November 2020, Granada, Spain*. By ed. Khalid S. Soliman, 1360 – 1369 (2020)
19. A.V. Bazilyuk. Financial motivators of economic growth and drivers of a new quality of life in Ukraine. The financial and credit mechanism of the development of the economy and the social sphere. *Materials of the II International Scientific and Practical Internet Conference, October 24-25, 2019*, Kropyvnytskyi: "Exclusive-System", 13 – 16 (2019)
20. O. Savruk, We don't have to "catch up" with the world - we have to stand by and do our part [online], Available at: https://tyzhden.ua/Economics/253984 (2021)
21. A guide to some aspects of conducting financial investigations (the investigation of financial relationship) [online], Available at: https://fiu.gov.ua/assets/userfiles/200/Typologies%20(National%20Studies,%20Guidances%20etc)/UKR_FINANCIAL%20INVESTIGATIONS%20MANUAL.pdf (2021)
22. P. Kutsyk, Cognitive approach to the formation of strategy based on DATA mining and SWOT analysis. Economic Analysis, 32(2), 243 – 251 (2022)

The impact of homeworking on work efficiency

Eva Kalinová[1,*] and *Lucie* Dolejší[1]

[1]Institute of Technology and Business in České Budějovice, School of Expertness and Valuation, Okruzni 517/10, 37001 Ceske Budejovice, Czech Republic

Abstract. The topic of homeworking has become very actual during the pandemic of covid-19. This pandemic showed the unpreparedness of companies for this modern way of working. At the same time, however, it contributed to rapid reaction and the introduction of homeworking into normal work processes. The aim of this paper is to find out to what extent and in which economic sectors the introduction of telecommuting has proved feasible. What benefits does it bring to employees and employers. And then to evaluate to what extent homeworking has an impact on the efficiency of employee performance. The analysis was based on data from the Czech Statistical Office's statistical survey as well as a survey conducted by the Ministry of Labour and Social Affairs. The data will be classified and evaluated using classification analysis. The results of this analysis have shown that the covid-19 pandemic has contributed to a faster uptake of home working even in organisations that did not use this way of working before. It also showed that the majority of employees interviewed would welcome a "hybrid" employment model in the future, where they would work from home 2-3 days a week and work in the office the remaining days. An evaluation of employee effectiveness yielded the finding that most managers were satisfied with the work-from-home arrangements in their organization. Based on this work, the use of homeworking can be recommended to some extent at some sectors in the future.
Keywords: homeworking, work from home, work outside the workplace, employee efficiency, work performance

1 Introduction

The concept of home office or working from home is currently a very actual and discussed topic. The use of this way of working has been largely driven by the COVID-19 disease pandemic. As a result of government regulations, companies have often been forced to reduce or suspend their operations. The Czech Republic and its neighbouring countries are currently struggling with the economic problems caused by the Covid-19 pandemic [1]. The business environment is determined by various obstacles the business owners have to deal with in the course of their work [2]. The COVID-19 pandemic has affected every aspect of consumer behaviour their expenses, investments, and financial reserves and social wellbeing [3].

*Corresponding author: kalinova@mail.vstecb.cz

Where possible, companies switched to remote working. This was often a completely new situation for many organisations, both for employees and managers. Everyone was forced to adapt quickly to the situation. Although remote working is now a very hot topic, it is not a new or unexplored topic.

Everyone was forced to adapt quickly to the situation. Although remote working is now a very hot topic, it is not a new or unexplored topic. Experts began to look more intensively at the subject as early as 1973 in the wake of the oil shock, when daily commuting became problematic. It was then that [4] came up with the idea that employees should not drive to work, but that work should follow them. Thus, the first "modern" motive for introducing distant working was to eliminate the costs associated with commuting [5]. However, in the following years, the motives for distant working changed. Other motives included, for example, the terrorist attacks on the World Trade Center on September 11, 2001, which caused people to worry about large concentrations of people in one place. With the rapid development of information technology, and therefore the improved availability of distant working, the importance of a better work-life balance for employees has begun to be emphasised. The covid-19 pandemic has again popularised teleworking but has not brought anything new to the field. All the reasons for homeworking were already there in the past [6, 7].

Some of the advantages and disadvantages are easily measurable, e.g., savings on the employer's side, while others are difficult to quantify. These depend on the employer-employee relationship and mutual trust. The advantages and disadvantages of working outside the employer's workplace have been discussed in detail. Publications also provide a search of research conducted on this topic in developed countries in the past decade [8].

The current analyses, therefore, contain themes that were already relevant before the pandemic, but now have become relevant to a much broader spectrum of workers. According to the results of previously conducted surveys, only approximately 4-8% of employees in the Czech Republic used homework before the covid-19 pandemic [8]. An important difference that the pandemic has brought about change in the perception of working from home.

The covid-19 pandemic was, however, perceived and offered as a benefit by the Home Office [9]. This benefit was used by companies to attract new or retain their permanent employees. Pandemic turned it into a necessity with regard to protecting the health of employees. Thus, telework has become relevant even for employees who do not want it. Based on the experience that companies have had over this, albeit short, period. It can be assumed that teleworking will become an opportunity for them to optimise the costs associated with running office buildings. [10] have conducted research that suggests workers and employers should avoid working from home if possible, even if it has a positive effect on the pandemic. It might seem that the company will save money with this form of work, but the reality is that the company's costs may even increase. This is due to the fact that most companies have chosen to work from home only partially, so the costs of running offices remain the same or at best slightly reduced. Legislation stipulates that the company must reimburse workers for the costs incurred by this form of work.

2 Literature research

The aim of paper is to investigate the impact of the covid-19 pandemic on the spread of distant working in different economic sectors. An important aspect, especially for employers, is the employee's job performance when teleworking and whether employees' performance declines or whether they are more efficient when working from home. Therefore, I will also focus on evaluating the productivity of employees working from home. In view of the stated objective of this paper, the following research questions are set:

Research Question 1: To what extent and in what economic sectors did employers allow Home Office during the pandemic. What has been the main reason why distant working has not been used more extensively?

Research Question 2: How do employees and managers evaluate the functioning and conditions for working from home?

The issue of distant working has been dealt with by many authors and researchers for a long time, although in the Czech Republic this topic has only become topical with the pandemic of covid-19. [11] who in their work "Flexible forms of work - homeworking in the Czech Republic and selected European countries" investigated the current state of application of one of the flexible forms of work. The performance of work outside the employer's workplace. In their work they present the main findings of the analysis of the application of this form of work, including its definition, identification of possible barriers to its advantages and disadvantages. The elaboration of this study was based on the use of secondary analysis of domestic and foreign information and data sources. Expert studies and outputs of national and international research projects carried out by the Research Institute of Labour and Social Affairs. Other research institutions whose focus is on the issue of flexible forms of work. Relevant statistical methods were used for a more detailed analysis and search for factors that may influence the use of home offices in the Czech Republic. According to the results of the [12] sample survey, 83% of employees perform their work at the employer's location, 2% usually work at home, 4% combine their work at different locations and 11% of employees have their work location forced by the nature and scope of their work. The results of the sample survey "New forms of employment" [12] show that most employees who have experience of homeworking work in small and micro enterprises. While large companies often offer an employee benefit in the form of homeworking. Only a small percentage of employees end up using this benefit. Small and micro businesses offer it less, but it is used to a greater extent. Enabling off-site work brings a range of impacts for employees and employers. According to [13], the impacts of off-site work for employers vary mainly by firm size or number of employees. For employers, the cost savings associated with running a workplace are usually a clear advantage, with small and micro enterprises with up to 10 employees being the most likely to mention this aspect in the guided interviews. For larger enterprises, working from home is particularly interesting as an employee benefit. For employees, time flexibility and reduced commuting are crucial. Yet time flexibility is not only emphasised by certain groups of employees for whom this is to be expected, but also for other employees who simply "want to be masters of their time" [13].

[14] examine the current state and future evolution of the workplace in the context of the covid-19 pandemic. They analyzed documents, publications, and surveys from multiple sources to learn more about employee experiences and the advantages and disadvantages of distant working. Documents from highly reputable sources such as Forbes, Gartner, McKinsey, Gallup, WHO, Boston Consulting Group, KPMG, PwC, EY, Deloitte and other publications produced between March 2020 and July 2020 were analysed. The methodology used therefore allows for a combination of all of these as the grounded theory allows for the collection of data from a wide range of sources. Czech Statistical Office provides data for analyzing process [15]. The data collected from the reports was categorised, validated and adjusted at all stages of the analysis. The data analysis showed how the workplace can be changed, re-vamped and five strategic categories were identified as part of this process:

1) To break free from the past.
2) Workplace redesign.
3) Technology.
4) Digital strategies.
5) Telework and a new sense of the workplace.

Ad a) While in the past organizations have only theorized about new forms of work, covid- 19 has forced them to move to entirely new ways of working [16]. Therefore, companies need to identify the most important processes within their organization, geographic and functional structure. New processes for the digital world need to be built, educated and prepared for the organisation so that there is no distinction between face-to-face and remote working. It will be necessary to review the economic policies of organisations in the area of remuneration and motivation. Future jobs will not be the same as those of the past, leading to the need to rethink performance measurement, and thus the evaluation will not necessarily be about the actual performance index, but about the employee's ability to overcome uncertainties and adapt to the ever-changing way of doing business.

Ad b) Offices and workspaces have so far been created without any real consideration of the underlying objectives or needs of staff. This too is something that Covid-19 has changed. Companies need to design spaces that can support all forms of interaction that take place on-site or remotely. A company that focuses for telecommuting will not require a large amount of office space as it could be replaced by home offices. Currently, we have eagerly been seizing the opportunity of online shopping. Buying goods via the internet is much more comfortable than hours wasted roaming through the shopping centre. After the economic crisis triggered by the onset of the Covid-19 pandemic, the online world penetrated all spheres of our lives [10].

Ad c) Technology helps in a collaborative work environment by enabling agile interactions in the workplace. Regardless of employees' past technological expertise, the pandemic presented an opportunity to improve the use of these tools. As demonstrated, the technology component played a critical role in telework and became a major survival factor that companies faced during the pandemic. For intra-company communication, it is advisable to use the social network Yammer, which is a suitable online social network for implementation in a small company [17].

Ad d) This category sees the acceleration of digital transformation as a trigger mechanism for telework and workplace transformation. And it also reveals important gaps within the IT infrastructure.

Ad e) Although most companies have been reticent to allow employees to work remotely, this pandemic has forced a worldwide experiment where many more people have had to work from home. What is remarkable is the speed and efficiency with which companies have adopted forms of digital collaboration, where in most cases the results have exceeded their expectations. One of the important factors for successful management is the knowledge of the competitive environment in which the business operates and the ability to adequately respond to changes [18].

Some of the benefits this has brought to employees and organisations include a reduction in transportation time and associated environmental impact, improved productivity and better time management, work-family balance coupled with the potential for greater happiness, an increased sense of protection and security in the face of potential pandemic impacts, the opportunity to attract talent without geographical boundaries, the introduction of new processes in line with remote working, strengthening company culture and reduced real estate costs [19]. When considering a future workplace, it is important to consider that remote working has clear benefits for people's wellbeing and workplace change. For future developments, for example, they conclude that allowing employees to work from a home office as a benefit to the company and employees in a less geographically constrained location. People will therefore live where they want, sometimes at a lower cost of living or close to their loved ones, which in turn turns the workplace into a more attractive place to work. The results show the existence of strategic categories that are important for re-imagining the workplace after covid-19 and for the process of changing it. These principles can be summarised as changing the physical space of the workplace, work flexibility, family reconciliation and health security [20, 3]. The study of these principles, based on data analysis, can open to different avenues of research, and may enable the creation of models that are directly related to employee satisfaction. Thus, it can be concluded that teleworking and the new ways of working that have emerged in the pandemic may even allow for greater workplace change than expected in the pre-pandemic situation. The workplace can be considered a friendly, familiar environment because it is set up in its own home. However, this does not negate negative feelings such as fear, uncertainty in case you must be at home for a longer period of time.

Factors influencing corporate social responsibility depend on basic competencies to influence corporate social responsibility.

Nowadays, human society is entering the era of digital economy with shortening product cycles and expanding scope of competition leads to increased competitive pressure [21].

For future developments, for example, they concluded that allowing employees to work from a home office as an advantage for the company and employees in less geographically constrained locations. People will therefore live where they want, sometimes at a lower cost of living or close to their loved ones, which in turn turns the workplace into a more attractive place to work. [22] say, that development of industrial environment and changes in technologies always brings negative influence on companies and employees, because of the fundamental skills and abilities.

3 Data and methods

It will be used document analysis to process the data. It will be proceeded in two steps. First, It will be found out which companies, in terms of size and industry distribution, allow their employees to work from home. It will be used data from the Czech Statistical Office, Use of Information and Communication Technologies in the Business Sector - 2021, Use of Teleworking. These data were based on the results of the annual statistical survey of the Czech Statistical Office on the use of information and communication technologies in the business sector in the Czech Republic. The reporting units were business entities with 10 or more employees in all sectoral sections except Agriculture, Mining and quarrying, Money and insurance and sections including Education, health and cultural activities. The sample was designed to ensure sufficiently representative results. The majority of the data were related to the first half of 2021. It will be then processed these data in Excel and will be used classification analysis to split the enterprises by size and by economic sector.

In the next step, it will be processed data based on the evaluation of the functioning of telework from the perspective of employers and employees. I will draw data from a survey conducted by the Evaluation Department, Section of European Funds and International Cooperation, Ministry of Labour and Social Affairs of the Czech Republic (September 2021). This survey was based solely on a questionnaire survey conducted in June and July 2021. Respondents were exclusively from the public and non-profit sectors. The following entities were contacted for the survey: organizational units of the state (hereinafter referred to as OSS), Territorial self-government units (hereinafter referred to as USC) and Private non-profit sector (hereinafter referred to as NGO). It will be processed these data in Excel. In this section, the number of respondents (the so-called "N") will be indicated for all processed graphs. The ambition of this elaboration will be to presented in an aggregated and structured way the information obtained from the respondents about their experiences related to working from home.

 https://doi.org/10.1051/shsconf/202316001004

4 Results

For the first part, the data from the CSO were used, based on the results of the annual statistical survey of the CSO on the use of information and communication technologies in the business sector in the Czech Republic. The structure of respondents is shown in Figures 1 and 2.

Fig. 1. Structure of the result set by size of firms (number of employees) in %.

Source: own elaboration based on CZSO statistics.

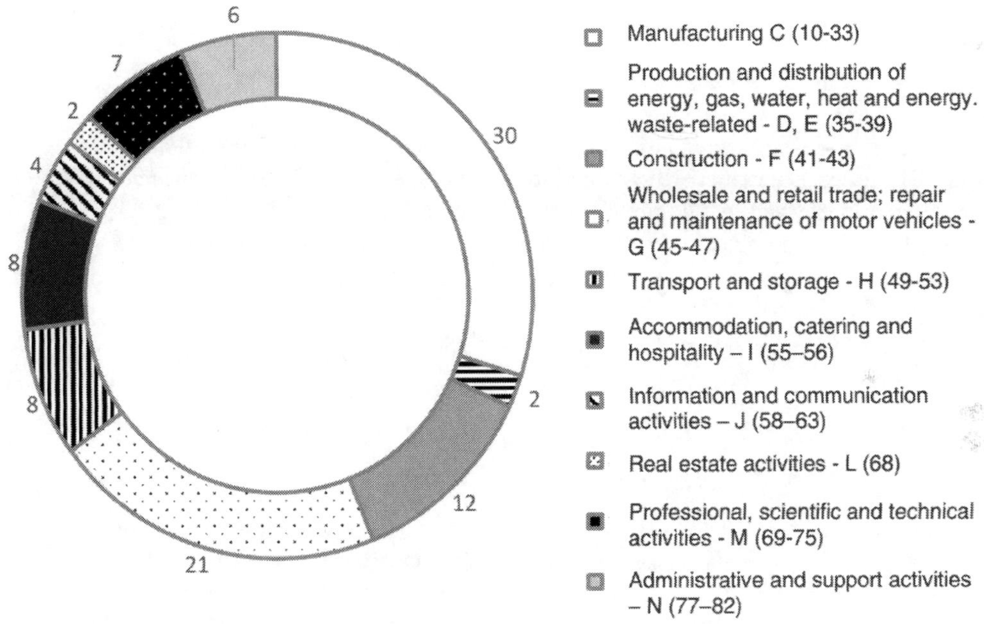

Fig. 2. Structure of the final population by sector.

Source: own elaboration based on CZSO statistics.

Main findings

In the period before the pandemic or during the loosened government measures, at least 54% of businesses sometimes allowed people to work from home. These options varied depending on the size of the businesses. While almost half of small firms (48%) and three-quarters of medium-sized firms (74%) were so, for large firms up to 91% could work from home. In terms of sector, firms in the information technology sector were the most likely to have their employees home office, followed by the audio-visual sector and travel agencies and bureaux. Some administrative workers in manufacturing also stayed at home (see Figure 4).

In contrast, 60% of companies allowed their employees to work from home during the pandemic. The home office option, however, varied significantly depending on the size of the company. At that time, 96% of large companies with more than 250 employees, 84 percent of medium-sized companies, and 53 percent of small companies with 10 to 49 employees allowed home offices. In terms of industry, the ranking did not change. A comparison of businesses allowing employees to work from home is shown in Figure 3.

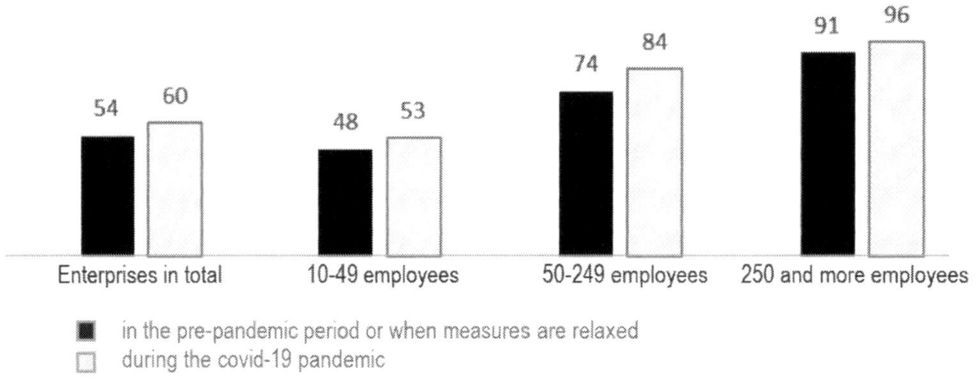

Fig. 3. Companies in the Czech Republic allowing employees to work from home (in percent).

Source: own elaboration based on CZSO statistics.

The difference, however, is in the end how many employees can actually work from home. During the pandemic, 12% of employees worked from home. According to the CSO data, teleworking was most common in large companies with 5 or more employees (21%). Teleworking was used most by the following sectors: telecommunications (69% of all employees in this sector), followed by employees in the media sector (59%), and employees in the IT sector (53%).

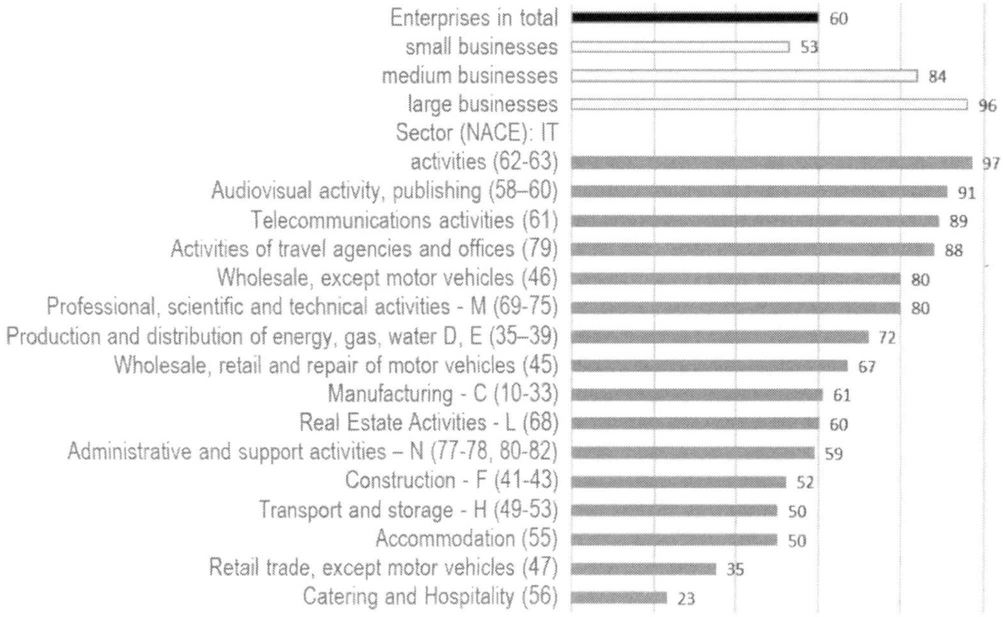

Fig. 4. Businesses that allowed employees to homeworking during the pandemic.

Source: own elaboration based on CZSO statistics.

In the second part were used data from a questionnaire survey conducted by the Evaluation Department, Section of European Funds and International Cooperation, Ministry of Labour and Social Affairs of the Czech Republic (September 2021). The distribution of respondents is shown in Figure 5 and Figure 6.

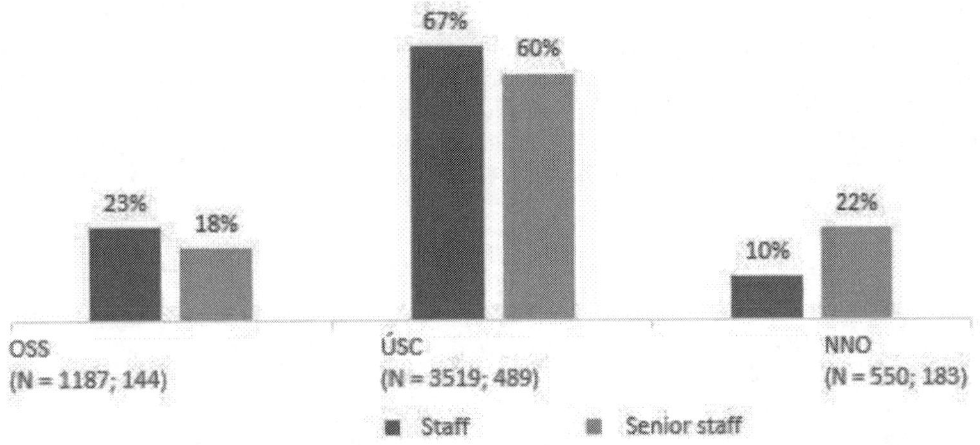

Fig. 5. Respondents by type of organisation.

Source: own elaboration based on the survey of the MLSA

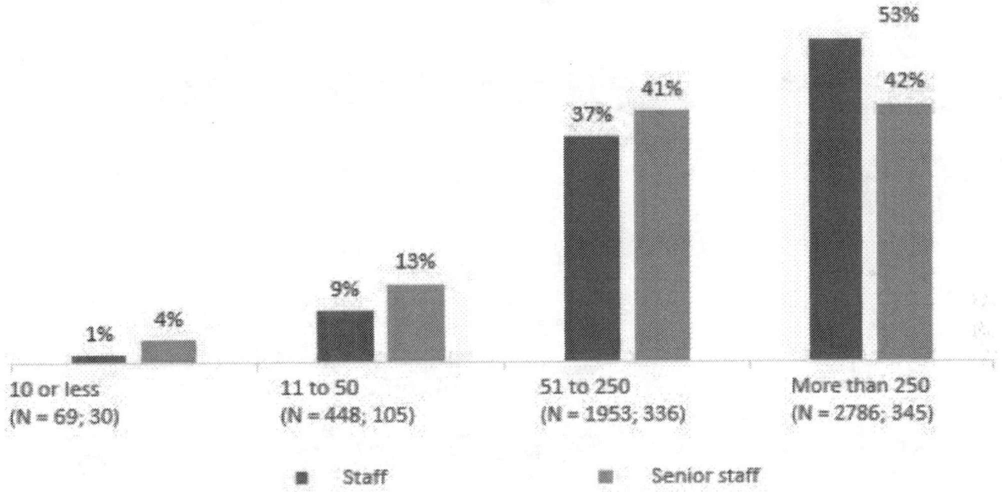

Fig. 6. Respondents by size of organisation.

Source: own elaboration based on the survey of the MLSA.

For comparison with the private sector, it has compiled an overview, based on a questionnaire survey of the MOLSA, of the situation of the possibility of working from home in the public sector in the period before the pandemic or during the loosened government measures and during the covid-19 pandemic.

The data evaluated was from the perspective of the employees. It shows that 49% of employees in the public sector and NGOs were allowed to work in the pre-pandemic period. These varieties depend on size and type of organization. The highest proportion in HO opportunities was allowed to employees in small organizations with up to 10 employees at 70% followed by large organizations with over 250 employees (65%). There was a major shift in the pandemic-19 related measures and a total of 87% of employees were allowed HO.

The largest shift in the possibility of HO was in medium-sized organisations of 51-250 employees, where 79% of employees were able to work from home during the covid-19 pandemic, up from an initial possibility of 26% before the pandemic. The distribution and differences are shown in Figure 7. In terms of the type of organisation, here the option to work from home is largely influenced by the work agenda that can be carried out from home. The questionnaire surveys show that 83% of OSS employees could work from home during the off-measure period, followed by 67% of NGO employees and, in contrast, only 32% of USC employees. These numbers increased during the pandemic-19 related measures. In the OSS, 99% of employees could stay at home, compared to 85% in the NGOs and 81% in the CSC. This was, however, an emergency situation which, after the end of the government measures back to the original numbers. The reasons for not being able to work from home are shown in Figure 8.

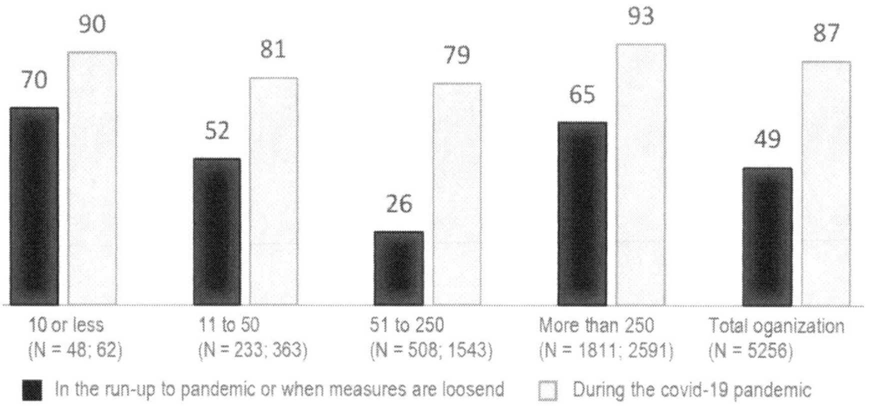

Fig. 7. The possibility of working from home in different types of organisations (in percent).
Source: own elaboration based on the survey of the MLSA.

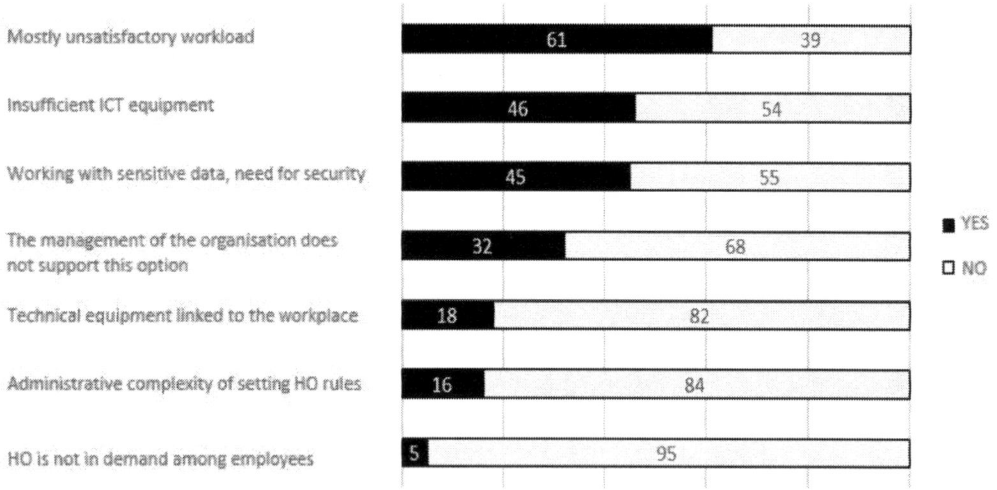

Fig. 8. Reasons why more employees do not use the HO option.
Source: own elaboration based on the survey of the MLSA.

Another result, which is the aim of this work, is the evaluation of the data obtained by the questionnaire survey of the Ministry of Labour and Social Affairs from the perspective of the evaluation of the functioning of working from home. First, it was evaluated the situation from the perspective of the managers. The vast majority of them are satisfied with the functioning of homeworking in their organisation. With the exception of the USC, only a minimal number of respondents are dissatisfied. In the case of the USC, this figure is 9% of managers. Overall, therefore, across sectors, over 94% of managers were satisfied or rather satisfied (see Figure 9).

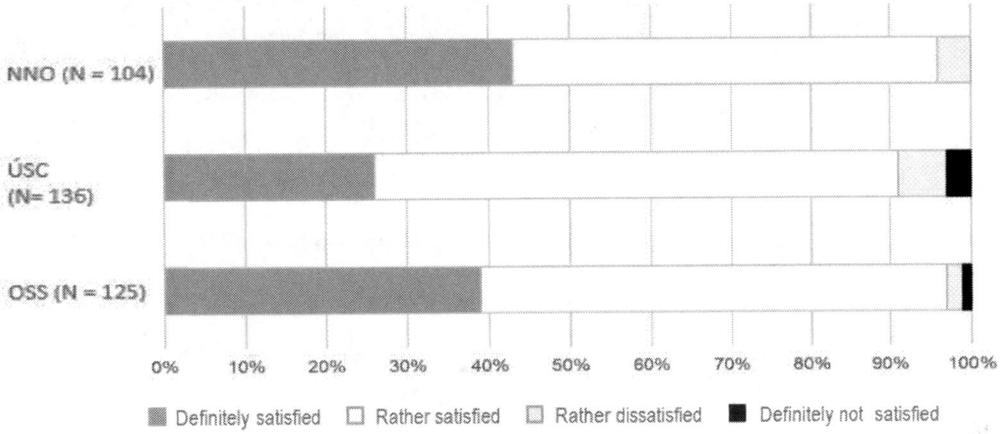

Fig. 9. Satisfaction of managers with the functioning of HO in their organisation.

Source: Own elaboration based on the MoLSA survey.

The results also showed a comparison of the performance of employees at the workplace and at HO (from the managers' perspective). The results show that the performance of employees working from home is comparable to working at the workplace (56% of respondents), with a further 11% of respondents perceiving employees working at HO as more efficient and only 8% perceiving employees working from home as less efficient. The remaining 25% of managers consider employee performance to be so individual that they are unable to assess it in aggregate (see Figure 10).

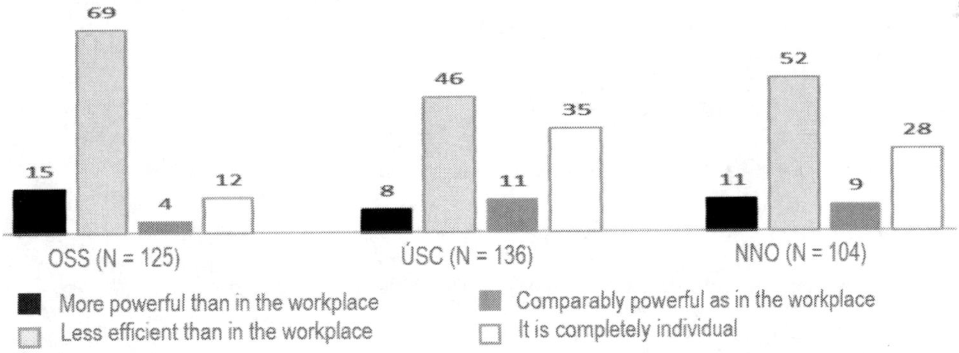

Fig. 10. Comparison of employee performance in the workplace and HO (in percent).

Source: own elaboration based on the survey of the MLSA.

From the employees' point of view, the questionnaire investigated how the respondents perceived working from home in terms of various criteria (efficiency, concentration, motivation, organisation of work in time, effectiveness of cooperation with colleagues, reliability of technology). Compared to working at the workplace, respondents rated the

 https://doi.org/10.1051/shsconf/202316001004

organisation of work time as the best, which was rated as significantly or rather better by almost 60% of respondents, with almost half of the respondents saying that they concentrated better and were more efficient at home. Respondents considered communication with colleagues to be the worst, with over 30% of respondents describing it as rather or significantly worse (see Figure 11).

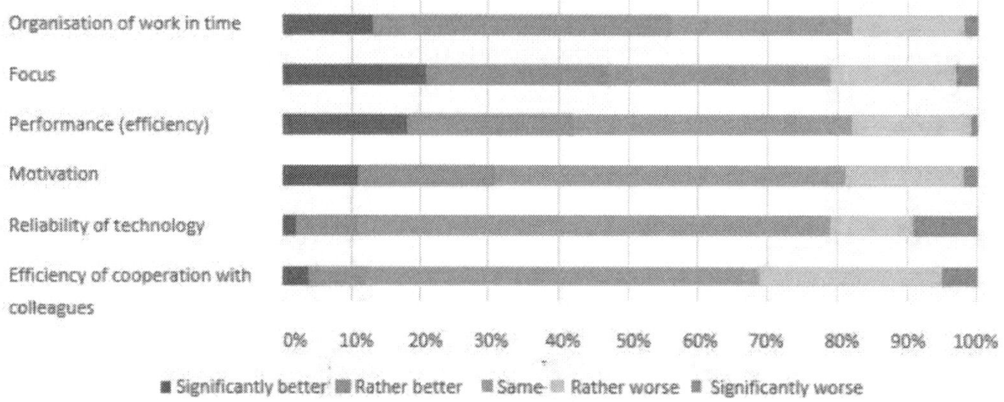

Fig. 11. Evaluation of the functioning of the HO compared to the workplace.

Source: own elaboration based on the survey of the MLSA.

The previous results are followed by the findings on how working from home affects employees' time management. A total of 60% of respondents said that working from home takes a comparable amount of time to working from the office, 20% of employees say that working from home takes less time, but these are mostly the ones who said they are more productive at HO. Conversely, about 10% of respondents need more time to complete their duties (see Figure 12).

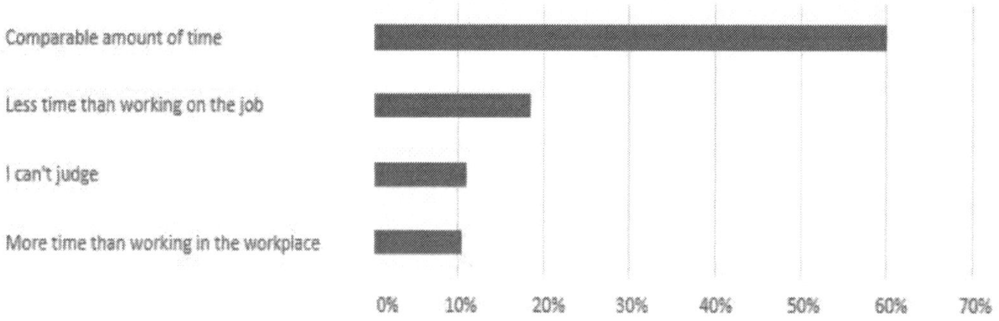

Fig. 12. Comparison of the amount of time required at HO and at the workplace.

Source: own elaboration based on the survey of the MLSA.

Based on the evaluation of the data available to me, I can answer the research questions. Question 1: To what extent and in what economic sectors did employers allow home offices during the pandemic? What was the main reason why telecommuting is not used more? Even before the pandemic, 54% of businesses in the country allowed employees to work from home. We find significant differences depending on the sector of economic activity. Employees of companies in the information technology sector were the most likely to work from home, while the smallest share was in the catering and hospitality sector or in the construction industry, i.e. in professions that cannot be performed from home. In the period of the covid-19 pandemic, the share of firms allowing home working increased further to 60%. The composition of economic activities remains the same, for the reasons mentioned above.

In contrast, based on the MoLSA's survey of public and non-profit sector respondents, only 49% of respondents were allowed to work in HO before the pandemic, but this number increased significantly above the private sector average during the covid-19 pandemic. Specifically, to 87%. This difference is due to the work agenda allowing more work from home. A largely unsatisfactory workload is also the main reason why organisations do not make use of HO options, a view expressed by over 60% of employees in the MOLSA survey. Another major barrier is the lack of IT facilities, with over 45% of respondents agreeing.

Question 2: How do employees and managers rate the functioning and conditions for working from home? According to the answers received from managers, it can be concluded that in most cases the management is satisfied with the overall functioning of telework in the organisation. Only 15% of managers expressed the view that employees are less efficient at home. In contrast, 10% of respondents said that employee performance is higher at home than at the workplace. However, the majority of managers (40%) were inclined to the option that employees are equally effective at home and at the workplace. Related to this are the brief conclusions regarding the most common control of work done from home. The survey showed that for most managers it is sufficient to monitor whether the tasks assigned are completed within the deadlines. From the employees' point of view, it is beneficial for them to work from home, especially with regard to the organisation of work overtime. This was identified as an advantage by almost 60% of employee respondents. Another almost half of the respondents said that they concentrate better and are more efficient at home.

5 Conclusions

The aim of paper was to find out the changes in the use of homeworking in the period before and during the covid-19 pandemic. And I also wanted to know the answer to the question of how homeworking works.

It was largely able to find the answers to these questions, although it is necessary to consider the structure of the respondents in both surveys. Based on this data, it was concluded that the covid-19 pandemic has contributed to a faster onset of enabling more companies and organizations to work from home. Given global developments, the expansion of the ability to work from home would have occurred gradually anyway. For companies, working from home means savings in operating costs. Although most employees do not want to use HO 100% there is the possibility of shared office space and therefore less need for large office buildings. Employees benefit from working from home, especially with regard to time management and also the time savings in commuting to work. But on the other hand, they lack social contacts when working from home for longer periods of time. For this reason, most employees would welcome a hybrid model where they would work from home about 2-3 days a week. As regards the functioning of teleworking and employee performance, here the surveys showed quite clear satisfaction of both managers and employees.

The stronger the core competence, the higher the corporate social responsibility will be, which means that the core competence can greatly improve the corporate social responsibility.

Based on these findings, it is likely that the use of HO will expand in the future, not only as an important employee benefit. Organisations will also start to consider more flexible office arrangements. Nevertheless, the results showed differences in the enabling of home working across different types of employers. These differences are due to the very nature of the work itself and the extent to which it can be done from home at all.

Even so, it can be assumed that the lessons learned during the covid-19 pandemic will permanently change the culture in many organisations, and the increased level of home working will persist to some extent after the pandemic crisis is over.

References

1. J. Horák, P. Dlouhý, Business risk in times of economic recession and economic growth. *Ad Alta-Journal of Interdisciplinary Research*, **11**(2), 82-86 (2021)
2. Z. Virglerova, L. Homolka, L. Smrčka, K. Lazanyi, T. Klieštik, Key determinants of the quality of business environment of Smes in the Czech Republic. *E & M Ekonomie a Management*, **20**(2), 87-101 (2017)
3. K. Valášková, P. Durana, P. Adamko, Changes in consumers' purchase patterns as a consequence of the COVID-19 pandemic. *Mathematics*, **9**(15), 1788 (2021)
4. J. M. Nilles, *Telecommunications-transportation tradeoff: Options for tomorrow*. John Wiley & Sons, Inc. (1976)
5. P. Mokhtarian, If telecommunication is such a good substitute for travel, why does congestion continue to get worse? *Transportation Letters*, **1**(1), 1-17 (2013)
6. O. D. Adekoya, T. A. Adisa, O. Aiyenitaju, Going forward: remote working in the post-COVID-19 era. *Employee Relations*, **44**(6), 1410-1427 (2022)
7. J. Vrchota, M. Maříková, P. Řehoř, Teleworking in small and medium enterprises (smes) before the onset of coronavirus crisis in the Czech Republic. *Management*, **25**(2), 151-164 (2020)
8. H. Vychová, R. Kyzlinková, S. Veverková, Working from home-description of the pre-pandemic situation and possible implications for the new conditions of work organization. *Social Policy Forum* (2020)
9. J. Srpová, *Starting a business: with case studies of budding entrepreneurs*. Prague. Expert (Grada) (2020)
10. J. Kučera, T. Krulický, P. Navrátilová, The trend of work from home and its advantages and disadvantages during the covid-19 pandemic: A comparative study. *Ad Alta-Journal of Interdisciplinary Research*, **11**(2), 145-150 (2021)
11. J. Kotíková, R. Kyzlinková, J. Váňova, S. Veverková, H. Vychová, *Flexible forms of work - homeworking in the Czech Republic and selected European countries*. Prague: VÚPSV (2020)
12. NFZ. *New forms of employment* [online]. Available on: https://www.eurofound.europa.eu/topic/new-forms-of-employment (2018)
13. R. Kyzlinková, P. Pojer, S. Veverková, *New Forms of Employment in the Czech Republic*. I. Edition. Prague: VÚPSV (2019)
14. A. L. Ancillo, M. T. V. Nunez, S. G. Gavrila, Workplace change within the COVID-19 context: a grounded theory approach. *Economic Research-Ekonomska Istraživanja*, **34**(1), 2297-2316 (2021)
15. CSO. Czech Statistical Office, *Use of information and communication technologies in the business sector - 2021* [online], Available at: https://www.czso.cz/csu/czso/vyuzivani-%20information-and-communication-technologies-in-the-business-sector-year-2020-current-month-2021 (2021)
16. J. Gifford, Remote working: unprecedented increase and a developing research agenda. *Human Resource Development International*, **25**(2), 105-113 (2022)

17. J. Kučera, P. Smolík, Analyzing functions and services in online tranding. *Ad Alta-Journal of Interdisciplinary Research,* **2**(1), 98-100 (2022)
18. O. Kutnohorská, J. Krištůfková, Application of Porter's five forces model to the Czech dairy industry after the abolition of milk quota. *Littera Scripta*, **12**(1), 1-14 (2019)
19. J. Kučera, M. Smolková, Online communication within a company: Case study of small company. *Littera Scripta,* **15**(1), 1-23 (2022)
20. MOLSA, Ministry of labour and social affairs. Working from home 2021. *Experiences of employees and managers* [online]. Available on: https://www.mpsv.cz/-/zamestnanci-na- home-office-are-at-least-equally-performing-as-in-office-work-from-home-is-on-the-rise (2021)
21. C. L. Sun, S. Skapa, J. Z. Liu, J. Horák, Y. I. N. Yang, Does core competence affect corporate social responsibility? *Journal of Competitiveness,* **13**(4), 132-150 (2021)
22. K. Kuba, F. Milichovský, Industry 4.0 and its impact on employees' age. *Littera Scripta,* **12**(1), 1-12 (2019)

The economic consequences of the war in the 21st century at the regional and global levels

Alla Kasych[1,*]

[1]Kyiv national university of technologies and design, Department of management and smart-innovation, Nemyrovych-Danchenko Street, 2, Kyiv 01011, Ukraine

Abstract. The article examines the theoretical and practical effects of wars in the 21st century on the economy. The concept of "military economy" is highlighted. The dynamics of military expenditures are analyzed. The reasons behind and types of wars that have occurred in the past and are currently occurring are made clear. Emphasis is placed on studying the economic consequences of Russia's war in Ukraine. The assessment of the potential of countries before the war testified and confirmed the aggressive goals of a large country that attacked a much smaller one. The war against Ukraine put the issue of preserving the independence of small countries on the agenda. The main focus of the essay is on examining the economic effects of the conflict, not just on the economies of the participating nations but also on the world economy. The instability of individual commodity markets, inflation, and migration became the main mechanisms of the war's influence. As a result, the indicators of economic growth of many countries of the world are deteriorating. The raised questions actualize the task before the world community of finding new methods of solving the problems of wars.
Keywords: war, economic consequences, military expenditures, global economy

1 Introduction

Russia's military aggression against Ukraine in the 21st century put on the agenda not only the security and existence of a multi-million European country but also actualized the task of researching the multifaceted effects of war on the development of individual countries and the global economy as a whole.

War is planned violence between governments that takes place over concerns of symbolic importance such as land, resources, and power [1].

The authors of the work [2] highlight that war continues to consist of the organized use of violence to achieve political goals. And in the 21st century, the war continues to be a real political event. The reasons of the wars are somewhat different: in Africa, they are local conflicts over access to resources like water and land; in the Middle East and between Russia and Ukraine - they are for geopolitical interests.

* Corresponding author: kasych.alla@gmail.com

Scientists [3] consider two approaches to understanding the causes of war: liberal and conservative, which have strategic consequences. Within the framework of the liberal approach, by promoting democratization, conditions are created for deterring aggression and resolving conflict. According to conservatives, democratization will fail and is not worth the expense. In addition, the authors consider active and passive wars. The authors conclude that with the advent of nuclear weapons in the 20th century, passive wars became more widespread.

The author of the work [4] studies hybrid wars. He concludes that in the future, in a hybrid war, opponents will use modern technologies, tools of the global economy, unpredictable tactics, asymmetric modes of operation, etc.

The authors of the paper [5] note that in the 1990s, there was hope that military power would be replaced by "soft" power. The author of paper [6] also concluded that the wars of the 21st century, in the vast majority of cases, will not be fought with the use of huge firepower and military capabilities. According to the study [7], the introduction of nuclear weapons fundamentally altered the global order, shifting the balance of power (often known as the "balance of terror") and restricting international conflicts to periphery regions.

However, this did not take place. And in the 21st century, wars are accompanied by ethnic cleansing and mass violence caused by great political differences.

According to the World Population Review, as of September 2022, military operations of one kind or another are taking place on the territory of 23 countries [8].

The most significant military conflict was Russia's war against Ukraine, the goal of which was defined not only as the destruction of the state but also the destruction of Ukrainians, taking the globe back to the parameters of World War II.

Population migration, the destruction of industrial and social infrastructure on the territory of Ukraine, the seizure of nuclear power plants, the displacement of commodity products on world markets - all these negative processes require research in the context of ensuring the sustainable development of both the country, the region as well as the global economy as a whole. With all its political and economic institutions, the civilized world does not always have the leverage to prevent or stop military aggression.

Understanding the consequences of all responsible persons should be based on complex analytical information. All of this calls for research on war models, its aftereffects, and steps to rebuild Ukraine.

Military conflicts in certain regions of the world have had global economic and social consequences. However, the most recent military battle is the biggest to occur in Europe since World War II and, for the first time, it aspires to entirely occupy a different nation.

The importance of such studies from an economic point of view is based on the need to form a complex of information about the negative effects for those politicians who are responsible for the start of the war.

The purpose of the article is to study the impact of the war of the 21st century from an economic point of view on the countries at war and the global economy in the context of developing economic tools for deterring aggression in the world. To achieve this goal, the following main tasks will be solved: generalization of the theoretical basis of the study of the impact of wars on economic and social processes; developing methods for analyzing the impact of war on national and global economies; determination of the main economic and social consequences of Russia's war against Ukraine; describing the key steps required to stop war or the effects of it.

1.1 Literature review

War as a research subject is in the field of interest of political scientists, sociologists, economists, and others.

The scientists who study wars more often focus on exploring the causes of aggression, which should allow influencing it to prevent it. However, no less important is the study of the economic consequences of the war, whose understanding will make it possible to formulate arguments in favor of preventing the war from the standpoint of preserving resources for development and not directing them to eliminate the consequences in the post-war period.

Scientific works devoted to the study of the consequences of the war should be grouped according to several main directions:

The first direction is the study of the nature and consequences of war in terms of their composition and types.

According to [9], a "war economy" is any form of economic activity, whether legal or not, that takes place during a conflict. However, the author, also establishes the existence during the war of three components:

the war economy, which includes the mobilization and use of resources to support a country's ability to wage war;

the shadow economy, which characterizes illegal activities and the use of war to obtain profits from them;

the coping economy - unites population groups that overcome wartime problems and survive.

Accordingly, the economy plays an important role in the ability of countries to wage war. However, it is more important to understand the consequences of war. Thus, the scientist [10] pointed out two systemic consequences of the war. Wars have had a significant impact on how the world functions today. Additionally, the likelihood of future battles is influenced by recent ones. Under such conditions, it is precisely the objective assessment of the consequences and the determination of sources of financing for the liquidation of losses that allows us to understand the scale of economic losses.

A slightly different approach to the classification of consequences is considered in [11]. Thus, it is noted that economic, criminal, and security (political) consequences arise as a result of the war. Another important conclusion made by the author: the end of the war is not the basis for the disappearance of problems.

According to [12] concludes that the long-term growth of civil activity is explained by the resilience of individuals and the scaling of these effects. These effects of the war can be attributed to population mobilization resources and a particular group's capacity for consolidation and activation.

The conclusions in the paper [13] provide some interesting points for discussion. The scientist concludes that the economic consequences of the war can only be either positive or negative.

The scientists [14] also indicate a mixed effect of war on GDP per capita, as the production of weapons and ammunition is counted as having a positive impact.

Among the most significant systemic consequences are a change in the economic environment and the cessation of the functioning of the economy for a long time. A change in the economic environment leads to a rapid recovery of economic growth after the war. However, the termination of business processes has a more negative effect than destroying physical capital. The scientists [14] note that the cost of war consists of three parts: the cost of resources used for war; destruction of physical and human capital; the decrease in GDP per capita measured during and after the war. The authors of the work [15] draw attention to the fact that war-ravaged societies are prone to humanitarian crises. Another significant drawback and potential catalyst for crime is the surplus of weapons left over from the war.

The second direction – territorial – involves revealing the specifics of the war in one or more countries. Undoubtedly, every war has certain features, that are important for determining the content of the measures necessary to minimize negative manifestations.

Thus, scientists studied the consequences of the war in Kosovo [11], Afghanistan [9], Vietnam [12], and others.

One of the last wars was the one in Syria. The paper [16] studied the economic consequences of the war in Syria, in particular, the economic and social consequences. Migration, poverty, and a lack of education for entire generations are just a few of the major effects that will negatively affect the future.

A territorial approach to the study of wars is important because each war has had regional consequences and significantly affects countries bordering the regions of hostilities.

The third direction is the global level of the causes and consequences of war. The scientists [17] take into consideration how wars are directly impacted by globalization. Scientists note that the interaction between globalization and war is complex.

The impact of the war on the world economy directly from the point of view of the impact on financial markets is studied in [18]. Scientists claim that international markets evaluate conflict scenarios and determine their severity. On this basis, the effects can be ambiguous: from positive expectations of the growth of individual sectors to a rapid adverse reaction. That is, the seriousness of conflicts and the level of their predictability determine the responses of the world financial market.

In the reverse direction, the question is raised in [19]. Scholars consider the "trade makes peace" thesis, which seems to be very logical since societies achieve economic benefits from trade, and wars disrupt trade.

The very process of globalization manifests itself not only in economic processes but also in technological and production processes. This may be globalization's beneficial effect on wars given its multidimensional nature.

Global integration, according to [18], increases cultural uniformity, which promotes world peace.

However, in some cases, globalization can even become a catalyst for conflict because global markets can facilitate weapons acquisition by individual countries.

The authors of the work [14] studied the impact of war on GDP per capita. The authors conclude that despite the war's positive effect on the development of the military-industrial complex, GDP falls due to the destruction of physical and human capital and lack of investment in physical and human capital.

According to scientists, conflicts are the primary barrier to the economic growth of low-income nations. In light of this, achieving the economic objective of closing the economic development indicators gap should become a barrier to the planning of new wars.

Based on the above, it is advisable to formulate the following main hypotheses: an objective assessment of the economic and social consequences of war can influence the duration of the war; small countries have the possibility to withstand wars in the 21st century.

1.2 Methodology of the research

The impact of war on economic and social processes has been studied in the works of scientists from slightly different positions. However, considering the outlined problem's complexity, we think it reasonable to use a comprehensive approach to studying the impact of war on economic and social processes.

The main methodological approaches, which were taken as a basis for developing the methodology of this study and were previously presented by scientists, related to the issues of the impact of war on: GDP dynamics [14], migration processes and poverty [16], opportunities for economic recovery [9]. The paper [20] examine war's direct costs (human and economic). The scientists [21] suggested returning to the concept of "wartime" as a phenomenon of the 21st century that impacts international relations and global politics. The relationship between time and war is stressed by academics, who also point out that little research has been done on the subject. After all, the longer the conflict, the greater the economic and social consequences.

The article's methodology is based on the fact that the study of the economic and social processes of the countries participating in the war is on the one hand extremely important and, on the other hand - quite difficult.

The importance of such an analysis lies in the fact that an objective assessment of the immediate and especially long-term consequences of war allows the leadership of countries and the world community to make timely decisions regarding the economic mechanisms of influence on the aggressor to prevent war. The analysis of war allows us to present its consequences in a systematic way in terms of characteristics that are important in the 21st century: the preservation of human life, development opportunities, and which were not important before - such as the conquest of someone else's territory. In order for society, even that of the aggressor country, to comprehend the effects of the war from the position of their country's growth and the perspective of future generations, emphasis on the economic and social costs should become a forceful argument.

The complexity of the research is because in the attacked country, there are real difficulties with operational monitoring of the situation, and the aggressor country closes data and tries to hide information about the actual state of the economy.

It is practical to research the conflict using operational and retrospective analysis techniques, the outcomes of which may be quickly used to create projections for the socioeconomic developments of the warring nations.

Operational analysis of the economy is important to determine how long the aggressor country will have the economic resources to continue the war. A retrospective analysis of the consequences of the war is important for the general public: and to establish the extent of the damage and determine the amount of reparations when the aggressor country wants to return to the system of international economic and legal relations; and to assess the prospective losses of the aggressor country.

The development of forecast trends in the development of countries after the war should demonstrate the long-term consequences of the war, which should become an additional economic argument both for the aggressor country and the global community regarding the use of anticipatory political measures to prevent an active war.

The study of the war in this article will be carried out according to the following main stages: analysis of costs in the military sphere, analysis of the military and economic potential of the countries participating in the war, analysis of the consequences of the war for the participating countries and the global economy.

1. Analysis of expenditures on the military sphere.

The military sphere is not only an important part of the global security system but also a driver of the development of the national economy. Considering the importance of military enterprises as workplaces, the importance of innovations and new technologies in the field of defense, which are carried out not only in the interests of business but also with significant government support, should be emphasized. On the other hand, ground-breaking discoveries made in the military help to address a number of social issues.

The following basic indicators are used to assess the importance that individual states attach to issues of defense through the development of the military sphere:

military spending in value terms and as a percentage of a gross domestic product;

state spending on defense research and development;

the dynamics of patenting in the field of weapons and military equipment worldwide and in individual countries.

Understanding future battles and technical prospects for ensuring global security require close attention to military breakthroughs and new technology.

2. An important stage of the analysis is **the analysis of the military and economic potential of the countries participating in the war**, which makes it possible to assess the duration and consequences of such a conflict. The nation's resources determine the outcome of each military conflict. Comparing the economic and military potential of the war participants allows: firstly, to predict the duration of the war and, secondly, to determine the methods of influence on the war-initiating country in the context of limiting the necessary resources. Assessment of economic potential requires analysis of the following indicators: territory size, population, total GDP and GDP per capita; GDP dynamics in recent years; the volume, dynamics, and commodity structure of the country's export and import; the level of globalization of the country's economy, among others. The country's military potential involves monitoring military expenditure, total military personnel, and air force personnel by category.

3. The analysis of the consequences of the war should be carried out at two levels: at the national levels of the countries participating in the war, as well as at the global level, taking into account the time horizon of monitoring - in terms of operational consequences and general consequences for the entire period of the war.

It is possible to predict the potential length of the war by dynamic (operational) monitoring of the chances that a country under military attack will survive and the chances that the aggressor country will keep waging war.

The following objects are included in the analysis's operative stage:

- losses of the state, which consist of actual direct losses from the destroyed infrastructure (property of the state, regions, business, civilian population;
- deferred losses of the state - losses due to a decrease in the country's human potential and potential GDP losses, losses due to a decrease in the country's investment attractiveness, state budget losses due to the need to increase unforeseen costs for one-time liquidation of the consequences of the war, etc.

Determination and valuation of the country's total losses for the entire war period are necessary to determine the amount of reparations after recognizing the losses incurred.

It is required to do an analysis for both nations:

- macroeconomic indicators (GDP size before and after the war, GDP dynamics, volumes, and dynamics of exports and imports);
- degree of economic globalization of the nation;
- demographic indicators (the number of the employed people before and after the war, the number of migrants, and the number of human lives lost);
- revenues and expenses of the state budget before the war and the size of the budget deficit during the war and after the war in order to determine the possibilities of balancing them.

Globalization, which has developed dynamically in recent years, has led to the synchronization of the functioning of world markets and the spread of negative manifestations to an increasing number of world countries. Regional wars have an impact on the economic and social processes of the respective regions, and individual wars have an impact on the global economy. That is why the beginning of hostilities requires research:

- economic process dynamics and business activity dynamics (GDP dynamics);
- the roles of the countries participating in the war in the global economy (export-import) in terms of individual commodity groups;
- circumstances of specific commodity categories on the global markets;
- monitoring of inflationary processes;
- environmental conditions;

- migration's dimensions and trends.

Separate consequences require developing and implementing a rapid response plan and a set of measures to support the country and its population. Restoring a country that has suffered aggression requires developing a large-scale, long-term plan with strategic goals, budgets for implementation, and substantiation of their funding sources.

Considering the difficulty of obtaining individual data, the analysis was carried out according to certain indicators presented above.

2 Results

Wars in the 21st century have significantly more significant negative effects than in the past. At the time of the mercantilists, the expediency of war was explained by the opportunities to increase the country's resources and gain advantages. However, in current circumstances, conflict is accompanied by the following from an economic perspective:

- firstly, by the destruction of significant resources previously invested in the development of the respective territories;
- second, the use of significant resources for the conduct of war that could be used to solve existing urgent human problems;
- thirdly, the emergence of significant global effects not only for the countries participating in the war but also for the global economy as a whole.

Thus, if we consider the recent wars in Syria and Afghanistan, the key effects for the countries were the fall in GDP, a decrease in the population's income level, and large-scale migration (Table 1).

Table 1. Macroeconomic indicators of Syria and Afghanistan.

Indicators	2010	2011	2012	2013	2014	2015	2016	2017	2018	2019	2020
Syria											
GDP, current US	252.5	73.6	44.12	22.55	22.0	17.62	12.45	16.34	21.45	22.52	18.81
GDP per capita growth, annual, %	4.4	4.2	-24	-23.1	-6.2	0.6	-2.8	1.4	2.4	5.2	-11.7
GDP per capita, US	11820	3050	2094	1039	1135	916	707	942	1268	1334	1075
Number of refugees, thousands of people	1	23	260	358	2272	3692	4566	4827	5479	5672	5556
Afghanistan											
GDP, current US	15.86	17.81	19.91	20.15	20.5	19.13	18.12	18.75	18.05	18.8	20.12
GDP per capita growth, annual, %	11.3	-2.7	9	2	-0.7	-1.6	-0.5	0.1	-1.2	1.5	-4.6
GDP per capita, US	526	512	557	569	565	556	553	553	546	555	520
Number of refugees, persons	6434	3011	16184	16861	300421	257553	59770	75927	72228	72227	72278

Source: [22].

The examples of the wars in Syria and Afghanistan show that despite significant human and material losses, civilized society cannot ensure their quick end.

The war in Ukraine opened a new page of military aggression and actualized the issue of redistributing the GDP of the world's countries to finance the defense industry.

Despite certain theoretical conclusions regarding the civilizational movement of society in the direction of reducing the number of wars, the global economy remains quite militarized. The primary patterns seen over the previous 20 years must be noted, taking into account metrics like overall military spending and its percentage of GDP (Fig. 1): the value of military expenditures is gradually increasing and, in 2021, exceed 2 trillion. dollars USA; the share of military spending in the value of global GDP gradually decreased from 5% in 1070 to 2.2% in 2021. From 2010 to 2020, the average annual growth rate of military spending was 1.99% against a 2.56% GDP growth rate.

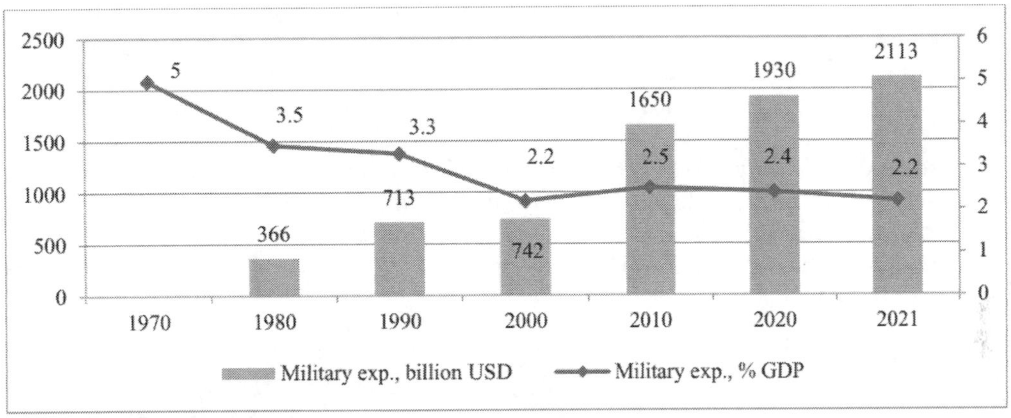

Fig. 1. World military expenditure.

Source: [22].

As noted in the paper [23], military expenditures in the coordinates of the "cannon-oil" concept no longer affect economic growth to such an extent that it should have become a factor in reducing such expenditures.

Most countries finance the army by 1-3% of GDP (Figure 2). The top spenders in terms of absolute amounts are the USA, China, India, and a number of EU nations. In terms of the share of GDP, Saudi Arabia is the leader.

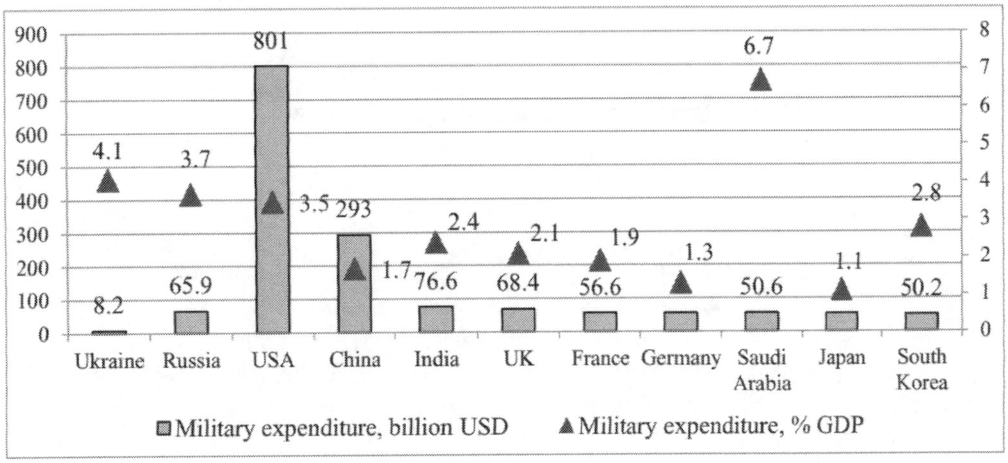

Fig. 2. Military expenditure several countries.

Source: [22].

However, the argument for increasing spending on the army, quite often, on the political plane and has nothing to do with economic expediency. Thus, in connection with Russia's war against Ukraine, the leadership of several countries announced an increase in spending on armaments next year. There was a change in the priorities of the planned budgets for the following years. Military spending, energy, and agriculture - these sectors have become priorities, while spending on education and social services will be reduced [24].

Countries with the most significant budgets for military development are the most likely to encourage the development of weapon innovations. The military advantage of the US is based on a powerful budget, the priority of which is spending on research and development both inside and outside the government. So, out of the 25 largest companies in the world for producing weapons and military equipment, 11 are from the USA, 7 are from China and 2 are from Great Britain. Total US investment in defense and military research (government plus business) still exceeds China's corresponding investment. However, China has tripled its basic science investment over the past decade. The existing trend will bring China to first place in the coming years.

Leading companies producing weapons and military equipment spend from 2-3 to 10-16% of their revenues on research, development, and similar activities.

Public spending on defense research in the United States is substantial and growing. So, in 2020, these costs increased: by 11.1% in the USA, by 11.8 in Australia, and by 3.9% in Germany (Fig. 3).

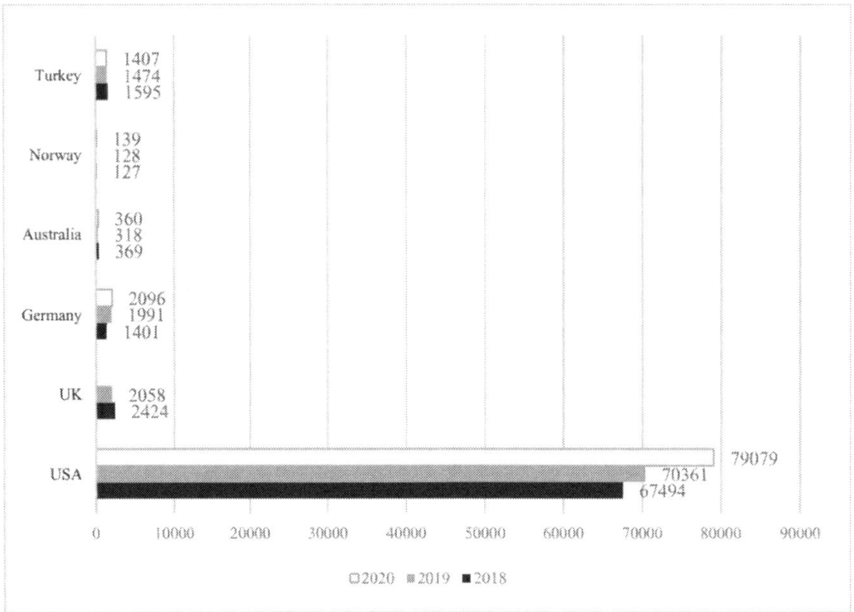

Fig. 3. State expenditures on defense research and development in individual OECD countries in 2018-2020, million dollars. US at actual PC prices.
Source: [25].

The relationship between the government and business, not only in terms of production and implementation but also in terms of scientific research, is the foundation of the international security environment in the OECD countries. Taking into account the current dynamics of military expenditures and expenditures for the development of science in the military sphere, it is safe to say that they will continue to grow.

2.1 Research of the economic potential of the war participant countries at the beginning of the war

The war in Europe in the 21st century was started by a country several times higher than the object of the attack by specific indicators.

European integration policies and the possibility for economic growth were factors in the conflict that Russia launched. This reason entirely agrees with the conclusion of the scientists [26] who noted that in a lawless environment, stronger states might fear that their security will be undermined by the economic growth of weaker states and may try to restrain it.

The primary indicators that provide a sense of the scope and outcomes of how Russia and Ukraine work show that the largest country in the world invaded a nation that, by some measures, is much smaller (table 2).

Table 2. Comparison of the economic potential of Ukraine and Russia.

Indicators	Ukraine	Russian Federation	Excess, times
The size of the territory, thousand km^2	603,7	17098	28.3
Population, million people	41	147	3.6
GDP, current billion US$, 2021	200	1780	8.9
Share in world GDP, %	0.21	1.85	8.8
GDP per capita, US$, 2021	2451	10219	4.2
Exports of goods and services, % of GDP, 2021	40.7	30.8	0.8
Imports of goods and services, % of GDP, 2021	41.9	21.3	0.5
KOF Globalization Index 2020 (place)	74.95 (44)	71.94 (49)	-
Military exp., billion USD (2021)	5.9	61.7	12.3
Real GDP growth at market prices in percent, %			
2021	3.4	4.7	-
2022	-45	-11.2	-
2023	0.8	0.6	-

Source: [22].

Russia has the largest territory in the world. The population of Russia is 3.6 times larger than the population of Ukraine. In other words, Russia is a much larger country that attacked a smaller country in terms of geographical size, potential for natural resources, human resources, and macroeconomic capacity. The most impressive is the excess of Russia in terms of military spending – 12.3 times. On the other hand, even the given macroeconomic indicators of Russia testify to the inefficiency of using the country's existing potential from the point of view of the logical strategic goals of the development of an ordinary country.

The economic potential of a country is the basis of its ability to create a gross product in the form of goods and services. The improvement of the key macroeconomic indicators and the standard of living of the populace serve as a measure of how effectively the economic capacity of the nation is being utilized. However, Russia's share of world GDP was 1.85% even before the war. And with the beginning of the war, there was a reorientation of the country's income to the resource support of the war.

Russia has a comparatively high degree of economic integration. Thus, the globalization index allows the country to take 49th place out of 196 countries worldwide. The share of exports of goods and services in GDP largely indicates the importance of world markets for developing the economy. The country's dependence on imports is quite significant.

Summarizing the prerequisites of the war that took place at the time of its announcement, it should be stated that the initiator of the war was a country that has powerful natural resources and economic potential, is fairly integrated into the world economy, but at the same time has a high level of militarization (including nuclear weapons). A positive answer was given to the rhetorical question of whether it is possible to fight with trading partners.

2.2 The impact of the war on Ukraine

Russia's unprecedented military actions led to: the occupation of 20% of the country's territory; loss of human lives and physical assets (property, housing, production, social and transport infrastructure); population migration; slowing down or stopping business processes; loss of labor force; relocation of enterprises from the east to the west.

The amount of one-time losses that Ukraine suffered from the Russian invasion is already 564.9 billion dollars [27]. Currently, the structure of destruction is as follows: 119 billion dollars. USA – infrastructure losses (destroyed and damaged almost 8 thousand km of roads, dozens of railway stations, and airports); 112 billion dollars. USA – GDP loss in 2022; 90.5 billion dollars. USA – loss of civilian population (10 million square meters of housing, 200,000 cars, food supply for 5 million people); 80 billion dollars USA - losses of enterprises and organizations; 54 billion dollars. USA – losses of direct investments in the Ukrainian economy; 48 billion dollars USA – state budget losses.

According to the latest data, at least 129900 residential buildings were destroyed and damaged, of which 114700 were private houses and 15100 were multi-apartment buildings. Direct losses due to the destruction of high-rise buildings amount to $42.3 billion, and another $5.4 billion in damages were caused to owners of private buildings. As of July 27, 2022, 2217 educational institutions and 903 medical institutions were destroyed [27].

According to the UN, more than 14 million people have left their homes; A large part of the population left the country, creating a challenge for Europe and the world in terms of large-scale migration. The countries that received the most Ukrainian migrants were Poland (1274 thousand), Germany (971 thousand), the Czech Republic (413 thousand), Italy (160 thousand), and Great Britain (112 thousand) [28].

In Ukraine, 745 enterprises have been relocated since the start of the relocation program, of which 558 have already resumed work.

The war contributed to a global food crisis. Numerous nations rely on wheat shipments from Ukraine, but since February, Russia has closed Ukrainian ports, sharply cutting grain exports.

The specified losses significantly impacted the macroeconomic situation, which is dynamically deteriorating. Only the most important signs of the war's widespread economic and social effects will be presented.

1. Dynamics of GDP. In recent years, the dynamics of Ukraine's GDP averaged 2-4% annually. The pandemic caused a deterioration in economic activity, but the drop in GDP in 2020 was not significant. The war instantly led to a decline in GDP growth rates, which in 2022 amounted to -15.1% (I quarter), -37.2% (II quarter), and -30.8 % (III quarter) (Fig. 4).

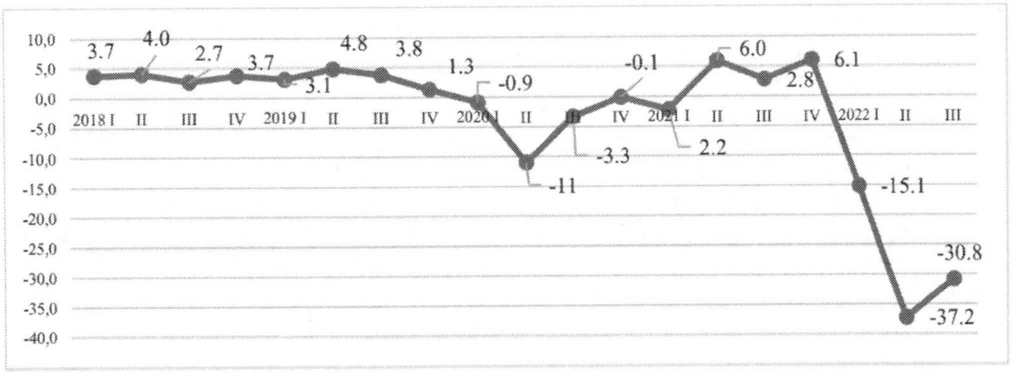

Fig. 4. GDP growth rate in Ukraine.

Source: [27].

*2. **Destabilization of the country's foreign economic activity.*** Significant changes occurred in the dynamics and structure of Ukraine's international trade as a result of the conflict, which resulted in the actual destruction of property or the blocking (sea ports) of many export-oriented businesses.). Dynamics of export and import (Fig. 5) of Ukraine by month in 2021 and 2022.

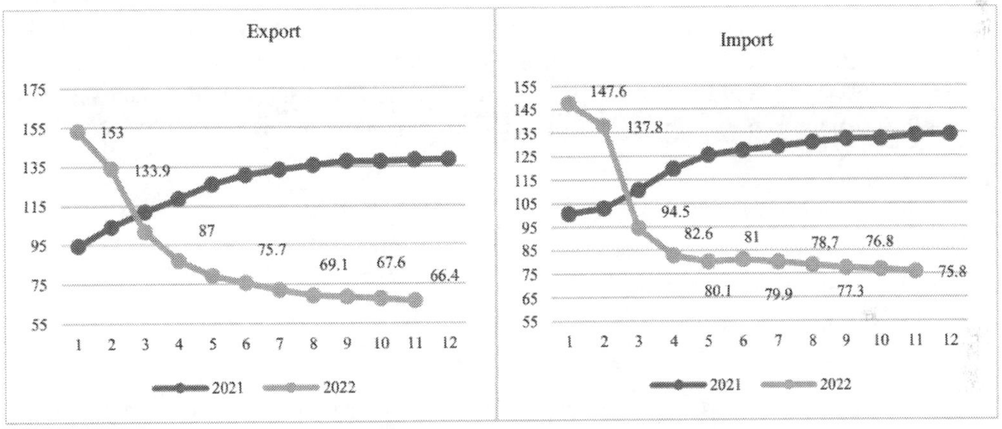

Fig. 5. Dynamics of export and import of goods, in % to the corresponding period of the previous year, cumulatively.

Source: [27].

As a result, Russia used food as a weapon at Ukrainian ports, giving the world an unprecedented example of its power. Because of the war, more than 20 million tons of grain intended for export, particularly for use within the UN World Food Program framework, were blocked.

*3. **Inflation for 2022 was 30%.*** A reduction in exports, a change in the commodity structure of imports, a drop in state budget revenues, and their reorientation to military purposes led to a revision of the national currency rate, which helped to stabilize foreign exchange reserves and maintain overall macroeconomic and financial stability. Let's analyze the pre-tax financial results of Ukrainian enterprises for January-September 2022. With the exception of financial and insurance operations, as well as the information and telecommunications sectors, we should draw attention to the losses of businesses in the great majority of economic sectors. This led to a significant decrease in state budget revenues. The deficit of the state budget of Ukraine needed to cover current expenses in 2022 amounted to 5 billion US dollars per month.

*4. **Significant transformations in the labor market.*** The physical destruction and relocation of enterprises and organizations, migration, and mobilization led to the loss of personnel potential by some enterprises and the forced reduction of employees by others. Overall, the unemployment rate rose to 40%, while in 2021, it did not exceed 10%.

Despite a certain adaptation of the economy to the conditions of the war in general, private consumption and investments remained very low due to the unprofitability of enterprises, disrupted logistics flows, and projected risks regarding the duration of the war.

2.3 The impact of the war and sanctions on the Russia

Russia's war against Ukraine raised several questions in politics and global security. However, from an economic point of view, the expediency of waging war with trading partners in conditions of a high level of internationalization of the economy is important and debatable. Russian territory was not only the largest, but it also played a significant role in world economic activity. And although Russia's GDP in the world GDP was only 1.75% before the war, the country was the world's largest exporter of fossil energy resources.

The recent pandemic-related macroeconomic dynamics (Fig. 6), which were based on the revenue from the export of energy carriers, were unstable. In recent years, Russia's share of fuel exports was more than 40-42% of GDP, while the share of high-tech exports did not exceed 9% of GDP. Revenues from oil and natural gas exports make up about 50 percent of Russia's federal budget. Before the war, Russia's economy could be characterized as export-dependent, energy-dependent, and import-dependent on high-tech products.

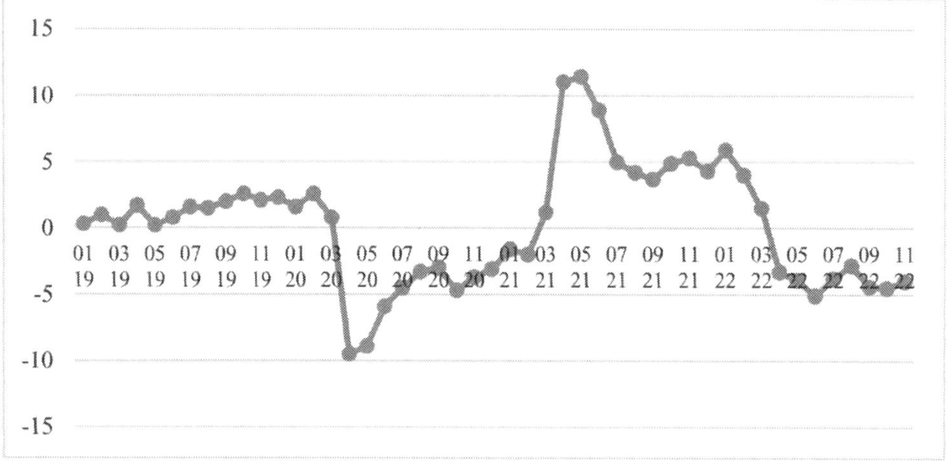

Fig. 6. GDP growth rate in Russia.

Source: [29].

In response to the military aggression, the world community developed and adopted a number of individual, corporate and sectoral sanctions aimed at using economic levers to end the war. The primary objective of the sanctions was to reduce Russia's capacity to fund the conflict.

The following should be mentioned in the lengthy list: blocking all Russian aircraft from entering EU airspace, forbidding financial transactions with the Russian Central Bank, imposing economic sanctions on the finance, energy, transport, and technology sectors, and prohibiting Russian banks from using SWIFT. The embargo on the import of oil and oil products, which will come into force in January 2023, should significantly limit Russia's income.

The financial system of Russia due to sanctions has already lost: 40% of market capitalization, 300 billion dollars of reserves of the Central Bank of the Russian Federation, 4.6 trillion rubles in the form of frozen Eurobonds, 10% of the capital of the banking system, 563 billion rubles of frozen retail assets, as well as assets held in depositories [29, 30].

Despite the economic sanctions, Russia's revenues from oil, gas and coal exports in the first six months of the full-scale war against Ukraine amounted to about 158 billion euros, far exceeding the costs of the invasion, which are estimated at 100 billion euros [29].

The Russian economy was able to adjust in some ways in the summer of 2022 after the first shock in February-March. The export of energy sources provided stable income, the decrease in import volume allowed to stabilize the exchange rate.

But the unfavorable patterns persisted through the end of the summer and into the fall. The sanctions began in mid-summer, as they had a limited effect and are only beginning to take effect. As a result of the fall in the production volumes of individual goods for July, they were: coal - 6%, gas - 25%, wood - 20%, plywood - 43%, cars - 81%, household appliances - 57%, tires - 43%, steel - 12 %.

Long-term systemic effects will arise in connection with the reduction of imports. Russia's import dependence by individual product groups in 2022 was: for goods of daily demand - 75%, sports goods - 48%, auto parts - 95%, toys - 92%, telecommunications equipment - 86%. High-tech import restrictions will cause a technological backwardness because machine-building companies are already reverting to outdated technologies.

The slowdown in the dynamics of economic growth occurs due to the termination of the world's leading companies from their activities in Russia. In total, 51 companies have finally left Russia, 1126 are leaving, and 491 are waiting. Such a state is accompanied by the limitation of investments and the implementation of their gradual withdrawal. All this indicates a gradual reduction in the demand for labor.

Table 3. Predictive indicators of Russia, %.

Indicators	2019	2020	2021	2022 (f)
GDP growth	2.2	-2.5	4.6	-11.2
Export	0.7	-4.1	3.2	-30.9
Import	3.1	-12.1	16.7	-35.2
Inflation rate	4.5	3.4	6.7	22.0
Budget Deficit	1.93	-4.0	0.72	-2.0

Source: [28, 30].

Against all the immediate and long-term effects, Russia continues to increase its war spending. Current costs for military payments in 2022 amount to 80 billion per month. These are payments without equipment, without clothing, and without the injured. The Russian Federation's draft budget for 2023 shows that defense spending will climb to $5 trillion from its baseline amount of 3.6 trillion rubles. However, Russia plans to hide the purpose of almost a quarter of its budget spending in 2023 and will significantly increase defense spending to wage a protracted war against Ukraine. The draft budget of Russia for 2023 foresees 6.5 trillion rubles, or 112 billion dollars, for classified or unspecified expenses.

War requires the use of resources not only for producing weapons but, above all, in the form of losses of human capital. The planned mobilization, which led to both the exodus of many young men from the country and their involvement in a considerable number of young men in the war, was a systematic error for the ongoing development of Russia (approximately 200,000 people). These processes lead to the withdrawal from the economy of the most productive part of the population this year. They will worsen the social structure of Russia in the long term.

When assessing the long-term consequences of the war in Russia on its economy, the following should be noted:

loss of export revenues due to economic sanctions of the world community and inability to take advantage of inclusion in global trade;

loss of export earnings as a result of international economic sanctions and failure to benefit from inclusion in global commerce;

decline of human capital as a result of intellectual mobility and mobilization;

differentiation of regions according to the level of economic development due to income inequality.

2.4 The impact of the war on the global economy

Russia's war against Ukraine significantly negatively impacts the global economy, which has already manifested itself in many economic processes and will have long-term secondary effects.

Fundamental mechanisms of the impact of war on the global economy:

1. Inflation was caused by the increase in energy and food prices.

Russia is one of the world's largest producers and exporters of oil and gas. The war immediately increased the risks of energy supplies from Russia. There was also a significant decrease in the volume of purchases of energy carriers due to the sanctions introduced by the EU and the USA, which caused a dynamic increase in gas and oil prices on world markets. Prices of agricultural products were added to energy prices. Ukraine exports sunflower oil – 42% share of global exports, maize – 16%, barley – 10%, and wheat – 9%. The blockade of Ukrainian ports led to an increase in grain prices as well. All these factors led to inflationary processes in many countries of the world.

Instability in the commodity market, which significantly impacts inflationary processes, will persist in the future. According to IMF experts, inflationary processes are expected in advanced economies at 5.7% in 2022 and 2.5% in 2023, in emerging and developing economies of 8.7% in 2022 and 6.5% in 2023 [31]. Inflationary processes are expected to affect bank interest rates and the monetary system.

2. Change in logistics flows in the energy market.

Russia is the second largest producer of gas and the third largest oil producer, accounting for approximately 17% and 12% of global production, respectively. The logistics of energy supply involves not only the creation of the appropriate infrastructure but also the consideration of political factors.

Limitation and price control of the export of energy carriers from Russia require effort and time. However, Russia began to use energy carriers, namely gas, as a weapon to blackmail EU countries regarding support for Ukraine. In 2021, Russia's share of natural gas imports to the EU was 39%. EU countries are phasing out Russian gas gradually. After the detonation of the Northern gas pipelines, it can be stated that, in general, the history of wars has shown an example of the use of economic levers for making political decisions on the part of the exporter. Currently, importers are changing logistics and trying to influence Russia's political decisions regarding the war.

Due to the sanctions, which provide not only the rejection of EU countries from energy carriers from Russia but also the impact on the prices of oil and gas from Russia, with the aim of reducing budget revenues and financial opportunities to continue the war. Revenues from oil and natural gas export make up 50% of Russia's budget. Incomes from oil alone amount to 10-15 million US dollars per month. During the war, Russia redirected oil supplies to China and India, increasing budget revenues.

3. Migration caused by war is a significant challenge for migrants and destination countries. The countries of the world, especially the EU countries, have experience in adapting to migrants. Still, these processes are always associated with several problems: creating conditions for simultaneous reception of a large number of people, carrying out their social adaptation and employment, and solving problems in the labor market.

The world had the largest migration flows in recent years due to the war from Syria - 6,761,000 people, and Afghanistan - 2,610,000 people. However, the Russian war caused the largest migration - more than 12 million people, of which 7,536 were relocated to the territory of European countries [28]. Solving migration problems, for example, from Ukraine, in the conditions of the destruction of their housing and property, is a significant challenge for both Ukraine and the host countries.

4. Global economic growth and GDP growth rates in EU countries began to slow down. According to IMF economists, global GDP growth would decrease from 6% in 2021 to 3.2% in 2022. In EU countries, growth rates will decrease to 3.1% compared to 5.3% in 2021, while European emerging market economies will slow to 0.9% compared to a significant growth rate of 6.5% in 2021 [31]. This applies to Poland, the Czech Republic, and Slovakia - the countries that receive the largest number of migrants and provide significant assistance to Ukraine.

3 Discussion

The research conducted from the standpoint of searching for optimistic scenarios for ending the war and overcoming its consequences allows us to express such warnings regarding the ongoing war, taking into account the discussion about the role of time in wars in the 21st century.

In 2022, Ukraine's economy faced a sharp drop in GDP and the large-scale destruction of its economic potential for further development. The nation therefore wants the war to end as quickly as possible. However, the war's end for Ukraine means the loss of sovereignty. And therefore, time could be better in priority to the goal of survival. To ensure the survival of Ukraine in 2022, the world community spent 28.1 billion US dollars. The cost of military aid is even greater. The fact that, for the first time since World War II, the world community is providing aid to Ukraine on a consolidated basis shows that a small country can survive and maintain its independence in the war of the 21st century.

The 21st century is characterized by the dynamic development of globalization processes, which leads to an increase in the level of internationalization of the economy of an individual country, including an aggressor country. The complexity of economic interactions between nations increases the degree to which nations throughout the world depend on trade in goods and services, technologies, etc. This should become an argument for decreasing interest in wars. However, for Russia, economic arguments not only did not become a safeguard but were also not taken into account at this stage of the war.

Therefore, an objective assessment of the war's economic and social consequences did not occur and therefore this is not an argument for preventing or suspending the war. The proposed theory was founded on the idea that social and economic issues matter more than political ones, which are not taken into account. The fact that Russia, after the entry into force of sanctions, that will significantly affect revenues, continues military aggression shows that the country can wage war for a long time, regardless of its economic consequences. And this question remains open.

In general, it is suggested that the following algorithm be used to do research on how conflict affects societal development:

1 stage. Study of the economic potential of the countries participating in the war. Such an analysis makes it possible to assess the ability of countries to support the conduct of war economically.

2 stage. Assessment of the consequences of war in various spheres. Negative effects are so multifaceted that they require a systematic study, the basis of which should be the impact classification. For the purpose of categorizing the effects of war, it is advised to apply the following criteria:

a) according to the time of manifestation of influence - some effects occur practically instantly, as well as those that have a long-term manifestation;

b) by the scale of influence – it is expedient to analyze the impact on the development of the countries participating in the war and the global economy;

c) according to the content of the impact following the concept of sustainable development, the war's consequences are felt in the social, economic, and environmental spheres.

Stage 3 – Evaluation of the mobilization resources of the war economy of the participating countries because the war economy is significantly different from a simple economy and requires reorientation according to the classic "guns-and-butter" model.

Stage 4 – Assessing the extent of damage to the country that has suffered aggression. This stage aims to determine the cost characteristics of the losses in the affected country.

Stage 5 – Analysis of the effects of sanctions of the civilized world relative to the aggressor country. The world community, in the context of limiting the economic resources of the aggressor country, implements a sanctions program, the effects of which are multifaceted and long-lasting, but forecasting these effects and understanding the nature of their action are necessary for future application against the aggressor countries.

Stage 6 – Determining the amount of resources needed to restore the country (according to the Marshall Plan model) that suffered from military aggression.

From the perspective of identifying causes and effects, the suggested algorithm can serve as the foundation for complicated war studies.

4 Conclusions

As before, the wars of the 21st century are aimed at solving political issues by military methods. A feature of the wars of the 21st century is the emergence of quick and long-term effects not only in the countries at war but also at the level of the global economy. In such circumstances, the militarization of the world continues to guarantee the ability of individual nations to preserve their independence, and nations who are unable to defend themselves militarily risk losing their sovereignty.

Ukraine faced threats to its existence as a state and the preservation of people's lives. The methods and goals of the war waged by Russia came from the 19th and 20th centuries. The economic logic of the 21st century, namely cooperation and globalization, is neglected. Using energy carriers, Russia was able to materialize threats against other nations, particularly EU nations. That is another feature of this war that the aggressor country used its exports as a tool for political blackmail of other countries.

The energy and food sectors of the global economy, and the financial sector, including inflation and the risks of sustainable economic growth, became sensitive areas that were affected by the war.

The ongoing war demonstrated an unprecedented attempt by the civilized countries of the world through economic sanctions to influence the aggressor country's political decisions and reduce the Russian economy's ability to finance military expenditures.

The following should be mentioned as some of the serious repercussions of the aggressor nation:

the loss of sales markets for a large number of products and the gradual loss of oil and gas export volumes, and therefore the gradual reduction of "energy" revenues;

restricting access to the nation's currency and gold reserves, which were stored in Europe and the USA;

a significant reduction in the import of a wide range of goods, among which there is a large share of high-tech products, which forms the trajectory of movement in the direction of technological backwardness, since the country is transitioning to technologies that were used earlier.

References

1. J. Levy, W. Thompson, *The Arc of War: Origins, Escalation and Transformation*. Chicago: University of Chicago Press (2011)
2. The Conduct of War in the 21st Century. Kinetic, Connected and Synthetic. Edited by Rob Johnson, Martijn Kitzen and Tim Sweijs First published (2021)
3. S. Metz, P. Cuccia, Defining war for the 21st century. *Strategic Studies Institute Annual Strategy Conference Report* [online], Available at: https://www.files.ethz.ch/isn/126626/pub1036.pdf (2011)
4. F. Hoffman, *Conflict in the 21st century: the rise of hybrid wars*. Potomac Institute for Policy Studies. [online], Available at: https://www.potomacinstitute.org/images/stories/publications/potomac_hybridwar_0108.pdf (2007)
5. C. Kennedy, T. Waldman, *Ways of War in the Twenty-First Century*. Issues in 21st Century World Politics (2017)
6. H. Munkler, *The wars of the 21st century*. IRRCMARCH, **85**, 849 (2003)
7. F. Coulomb, J. Dunne J. Economics, conflict and war. *Real-World Economics Review*, Peacon, 147-157 (2008)
8. World Population Review [online], Available at: https://worldpopulationreview.com/
9. J. Goodhand, From war economy to peace economy? Reconstruction and state building in Afghanistan. *Journal of International Affairs*, **58**, 1, 155-174 (2004)
10. M. Thompson, *The White War: Life and Death on the Italian Front, 1915-1919*. Faber & Faber, 496 p (2008)
11. J. Peterson, *Transforming a war economy: learning from the case of Kosovo*. Building a peace economy? (2018)
12. J. Barcelo, The long-term effects of war exposure on civic engagement. *Proceedings of the national Academy of Sciences*, 188 (2018)
13. G. Schein, *Invisible War*. World Literature Today **91**, 23 p. (2017)
14. C. Thies, C. Baum, *The Effect of War on Economic Growth*. [online], Available at: https://www.cato.org/cato-journal/winter-2020/effect-war-economic-growth (2020)
15. M. Marshall, G. Elzinga-Marshall *Global Report* Conflict, governance, and state fragility. Center for Systemic Peace Vienna, VA USA, 60 p. (2017)
16. War in the Region 2022 International Bank for Reconstruction and Development. The World Bank [online], Available at: https://www.worldbank.org/en/home
17. A. Tidwell, C. Lerche, Globalization and conflict resolution. *International Journal of Peace Studies*, **9**, pp. 47-59 (2004)

18. G. Schneider, V. Troeger, War and the world economy: Stock market reactions to international conflicts. *Journal of Conflict Resolution* **50**, 623–45 (2006)
19. C. H. Anderton, N. Beck, J. R. Carter, H. Dorussen, E. Gartzke, R. Gissinger, H. Hegre, Q. Li, E. D. Mansfield, J. D. Morrow, J. R. Oneal, J. C. Pevehouse, S. W. Polacheck, R. Reuveny, B. Russett, G. G. Schulze, E. Weede. *Globalization and armed conflict*. Rowman & Littlefield, 365 p. (2003)
20. S. Kushi, M. Toft, Introducing the military intervention project: A new dataset on US military interventions, *Journal of Conflict Resolution* **1776.** DOI: 10.1177/00220027221117546 (2019)
21. A. Hom, L. Campbell, Wartime in the 21st century. *International Relations* **36**(4). https://doi.org/10.1177/00471178221134 (2022)
22. Indicators [online], Available at: https://data.worldbank.org/indicator
23. S. Awaworyi, S. Ling Yew, *The Effect of Military expenditure on growth: An empirical synthesis*. Discussion paper 25/14. Monash University (2014)
24. T. Deen, *Russian invasion of Ukraine triggers a rise in military spending in Europe*. [online], Available at: https://www.indepthnews.net/index.php (2022)
25. Government budget allocations for R&D. [online], Available at: https://stats.oecd.org/Index.aspx?DataSetCode=GBARD_NABS2007# (2022)
26. N. Monteiro, A. Debs, An economic theory of war. *The Journal of Politics* **82**(1) (2020)
27. Refugees from Ukraine [online], Available at: https://data.unhcr.org/en/situations/ukraine
28. Official website of the State Statistics Service of Ukraine. [online], Available at: https://www.ukrstat.gov.ua/
29. Russland nimmt mehr mit Energie ein, als es für den Krieg ausgib[online], Available at: https://www.spiegel.de/wirtschaft/russland-nimmt-mehr-mit-fossilen-brennstoffen-ein-als-es-fuer-krieg-ausgibt-a-172f382a-b85d-437e-992c-7b005dd6b31d
30. J. Sonnenfeld, S.Tian, F. Sokolowski, M. Wyrebkowski, M. Kasprowicz, *Business retreats and sanctions are crippling the Russian economy* [online], Available at: https://ssrn.com/abstract=4167193 or http://dx.doi.org/10.2139/ssrn.4167193 (2022)
31. Global prospect and policies. [online], Available at: https://www.imf.org/-/media/Files/Publications/WEO/2022/October/English/ch1.ashx (2022)

What asset structure generates the highest possible profit for a manufacturing enterprise?

Tomáš Krulický[1,*] and *Lucie* Nováková[1]

[1]Institute of Technology and Business in České Budějovice, School of Expertness and Valuation, Okruzni 517/10, 37001 Ceske Budejovice, Czech Republic

Abstract. The objective of the paper is to determine the most suitable operational redundant assets and asset structure for a manufacturing enterprise. They are determined by the financial analysis of ratios, vertical and horizontal analysis, and the calculation of Altman Z-score. The recommended sizes of activity ratios are: the turnover of total assets should be equal to the value of 1, the time of inventory turnover should be as short as possible, the turnover of inventories should be as large as possible, the time of debt collection should be the shortest possible, the due date of fulfilling obligations should be equal to 2 or higher. The ratios are compared with the sector and the position of enterprise is assessed. The limit of the research is the incapability to determine the best asset structure, but only the best possible one, on the basis of comparing with the sector that sets the bar in a specific field. Every enterprise aspiring to be competitive should meet the sector standards, at least.
Keywords: asset structure, financial analysis, activity ratio, profitability ratio, liquidity ratio, comparison with the sector

1 Introduction

Using all the asset structures available an enterprise gains a competitive advantage directly inside the business. It focuses on the nature of resources, the way of efficient use, flexibility of resources and its balance in the overall corporate [1].

The resources of assets must be defined and financially evaluated first. We follow the principles of correct asset accounting. It is important to avoid manipulating balance sheets, artificial increasing their liquidity and altering the structure of assets [2].

It is necessary to consider what asset structure prevails in the enterprise and what resources fund it. A high proportion of debt capital results in a negative profit, the loss of enterprise. The optimal share of debt capital maximising the return on assets oscillates within the range of 0 to 21% [3]. The asset structure is bound to the scope of business, therefore, I intend to deal with an industrial sector where current assets prevail. Industry 4.0 integrates smart and connected production systems that are pivotal in predicting and supporting production in real-time, leading to sustainable organizational performance. In manufacturing,

* Corresponding author: krulicky@mail.vstecb.cz

it may increase productivity, sustainability, and energy efficiency, while optimizing competitiveness [4].

Costs can be minimized by correct optimising the structure of volume of used assets that is operationally necessary for the scope of business [5]. The asset structure of enterprise also depends on technological complexity of production – it especially affects the proportion of tangible and intangible assets, investment assets, the economic situation of enterprise and its ability to innovate the assets. An asset structure can be found in the balance sheet, on the side of assets. Profitability is one of the most important ratios of company activities. The recommended methodology of complex profitability analysis can be successfully applied to the companies of various sizes and activities [6].

Every enterprise must flexibly respond to the optimisation of its asset structure to prevent the deficiency or surplus of asset structure. The aim of this work is to determine the optimal asset structure in a manufacturing enterprise.

The following research questions were formulated to meet the aim:

VO1: How to determine redundant assets?

VO2: What asset structure is the most appropriate for an enterprise?

2 Literature research

[7] develops a framework within which the losses of the effect on the future performance of a company are used for the development of three categories of capital and he suggests a measure for each category. He combines these categories with the qualities of assets into which the capital should be invested and which involve both the aspects of liquidity and the resources of this capital. The procedures are based on three following principles: 1) the allocation of capital based on accountancy; 2) the allocation of capital according to risk; 3) incentive capital allocation. Considering the sequence of stages in the model of asset optimization structure together with the methodological aspects of determining the efficient structure of assets – maximizing a market value and profit, [8] suggest an algorithm for creating an optimal structure of assets. It is developed on the basis of financial statement with the structures of assets by carrying out vertical analysis of balance sheet asset and placing the individual assets on the basis of weight provided it assumes more than 1% of total assets. However, if it takes into account the composition of inventory involving the stocks of raw materials and other materials, unfinished production, completed products and commodities, the methodology of panel data is used [9]. The economic results of a company are an important tool for many entities, e.g., for internal entities as well as for external entities. As the economic results of a company are often the only source of information that informs the company's partners about the managerial activities of their company, it is necessary to present these economic results using real numbers. However, companies prefer to achieve better results by applying the principles of creative accounting, which leads to improved economic values being shown to be achieved during an accounting period [10].

As shown by the analysis, there occurred a decrease of the proportion of inventory on the total assets and current assets. The realisation of BMI business value requires a concentrated effort of the asset owner so that the owner could determine and assess critical activities controlling the business value in AM. [11] adopted the approach of qualitative research based on the strategy of multiple case studies the aim of which was to identify the key trade processes supporting the business value of BMI in AM. The study identified the six systems of critical activity that increase the business value of BIM for the owner of assets, i.e., BIM strategy, contract management, lifecycle management, maintenance management, order management and value realization management. The balance of asset model is depicted by [12] as a structure of class where the effect is sorted according to various types of assets. He proposes a theory of endogenous differences in the liquidity of assets based on an interaction between the differences in their risk and the differences in liquidity needs.

Not only do companies with a lower leverage effect have a lower probability of financial distress, but they also have a better position to obtain assets from the others in distress. [13] claim that using certain debt capital has an effect on the size of the profit of enterprise. The leverage effect can render companies reduce their debt as soon as the assets become more transferable. Therefore, it only has an effect to decrease debt. How do enterprises respond to asset structure in a crisis period? By a comparative analysis, [14] demonstrate that enterprises mainly survive during crises by relying upon the internal resources of financing. An external leverage effect is lowered as the costs of debt increase the risk of failure. In need, enterprises sell out insufficiently used or redundant tangible assets as they focus on their main competences and develop intangible assets. The management of stable development of business entities requires improving methodological approaches concerning the assessment of their efficiency. [15] assessed the efficiency with a comparative analysis with other economic entities regarding the efficiency of using the current number of resources. A classical approach to the analysis of efficiency by the means of the profitability of ratio do not always result in objective outcomes and require considering various economic and production factors. In practice, it is important to assess the efficiency of business entities by the means of a methodology involving two aspects. The first aspect enables to calculate certain indicators of relative efficiency characterising a specific enterprise. The second aspect is based on carrying out a comparative analysis with other economic entities concerning the efficiency of using current material, financial and human resources.

However, the enterprise value must also be analyzed on the basis of other key indicators, such as financial leverage, the Capital Asset Pricing Model (CAPM) method, or the net present and future value of the FCFE. This is especially important so that the results can be put into mutual relations and a sufficient representative value of the FCFE results can be achieved [16]. Value generators can be used to determine which asset structure is most profitable. They indicate the activities and capabilities that increase profitability, reduce risk and promote company growth. However, the main assumption is that the value of the company will be measured by the EVA Equity indicator [17].

[18] show, theoretically and with the help of real data, Bayesian efficient limit extends the efficient limit of sample; current estimator of the set of optimal portfolios that is known to be excessively optimistic. It also provides a back predictive distribution of optimal portfolio return that can be used for the construction of prediction interval. Why is financial analysis an important instrument for proving and controlling decisions on using assets? Since the turnover and return on assets substantially depend on the particularities of a sector and conditions for entrepreneurship, the means of financial analysis must adapt to specific situations, [1] claim. A methodological base is a systemic approach to dealing with the analytical management of assets. In the course of generalising the methodological principles of financial analysis addressing the efficiency of asset management, the techniques of abstract-logical methods are employed, i.e., analysis and synthesis, induction and deduction, analogy and comparison. The methods of average and relative values, the methods of modelling and factor analysis were used for all calculations; they were adopted for establishing a relation between the systems of financial ratios of asset efficiency. [19] critically consider the performance of individual sector on the basis of totally monitored indicators in comparison with a performance indicator (PI) and a key performance indicator (KPI). The paper contains a design of KPI system that would enable to assess the performance of industrial sectors including a correlation analysis of these indicators for long-term relations. Therefore, it is possible to assess the performance of individual industrial sectors with their help. These indicators are derived from total indicators that are monitored by the Ministry of Industry and Trade and the Czech Statistical Office [20] for rating development, manufacturing and industry. The methodology of a complex analysis recommended by [21] includes: 1) setting the aims of

analysis; 2) the selection of analytical resources and technical methods; 3) the calculation and assessment of profitability of sales (gross and net), profitability of assets (total assets, fixed and short-term assets), profitability of equity (share capital, equity and capital employed) 4) determination of factors influencing profitability; 5) use of information collected during the analysis for the purposes of internal and external users. [22] designed an indicator estimating the financial situation of an organisation; several publications directly regard Z as a standard for the risk of insolvency. Altman Z-score estimating the risk of insolvency fails to explain the historical return of assets.

The most suitable methods for fulfilling the aim of the work seem a financial analysis, assessing the ratios of assets and Altman Z-score.

3 Materials and methods

3. 1 Data

The source of data are financial statements of Izos, limited liability company, in manufacturing industry, for the years of 2016, 2017 and 2018. The data are to be analysed by the table system of Microsoft Excel. The fundamental source of data are the annual reports to be subsequently published at https://www.justice.cz/web/msp [online]. [quote 2022-04-10]. The financial statement includes a balance sheet, a profit and loss, cash flow and attachments. The following items of the balance sheet will be used: working capital, total assets, inventories, debts, short - term financial assets, current assets, equity, bonds and bills, current account, short - term liabilities, held profit, debt, short - term debt, quick assets. The profit and loss will provide the following items: sales of goods sold, costs of goods sold, performance, production consumption, personnel costs, depreciation, interest expense, HV for the accounting period, total revenues, total costs, earnings before interest and taxes, sales. The items will be compared with manufacturing sector, the data will be provided at: https://www.mpo.cz/ [online]. [quote 2022-04-13]. The financial analysis of manufacturing industry will be used for a precise comparison with Izos, limited liability company.

3. 2 Methods

The items for the years of 2016-2018 will be compared and analysed by financial, horizontal and vertical analysis. Financial analysis is an instrument for an objective assessment of enterprise; individual pieces of information, and their interaction, are assessed by the horizontal analysis of data in the course of time; the structure of an enterprise is assessed by vertical analysis of data. During analysing the assets of balance sheet vertical analysis provides the answers for questions, what percentage of total assets account for the individual asset groups (intangible assets, tangible assets, current assets, or accruals).

Aggregated items may be graduated if there is an interest in closer detail. We can determine what percentage of tangible assets are movable and immovable things, what percentage of current asset is its every item. What measures can be derived from vertical analysis? For example, to enhance the control of costs, to achieve increased profit for the following year or to prevent its decrease in comparison with the past, at least.

The individual items of assets in the balance sheet will be measured against total assets by vertical analysis. We will use it for calculating the share of individual items of assets in the total assets of enterprise in percentage. We will measure the difference of the previous year with the following year, in relative units, by horizontal analysis.

We will monitor the different values of an item in the period of t and **t+n** by the total growth of item value. The percentage change in the items of outputs should generally be higher than in the items of inputs. Thanks to horizontal analysis we identify the items with the biggest relative change. On its basis, it is possible to determine the degree of reproduction and the use of enterprise assets.

Horizontal analysis is calculated as follows:

$$total\ size\ of\ change\ =\ financial\ ratio_{\ t+n}\ -\ financial\ ratio_{t},\qquad (1)$$

where t means the basic monitored period,
t+n immediately following period.

We will examine the items from the balance sheet: fixed assets and current assets.

Vertical analysis is calculated:

$$The\ percentage\ ratio\ of\ certain\ item\ =\ X\ /\ total\ assets\ *\ 100\qquad (2)$$

where X means a certain item from the balance sheet.

We will examine the items from the balance sheet: fixed assets and current assets.

Financial ratios enable to acquire a quick picture of financial characteristics of enterprise. We will express the relation of two items from financial statements by their ratio. We will examine the following ratios of financial analysis for our needs: profitability ratios, activity ratios and liquidity ratios.

Activity ratios

They include: the turnover of assets (AR), inventory turnover (OZ). Activity ratios provide information how effectively an enterprise can appreciate its assets.

We will examine the following items from the balance sheet: total assets, inventories, debts; items from profit and loss: sales.

The turnover of assets is calculated from the relation:

$$OA\ =\ T\ /\ A\qquad (3)$$

where OA means the turnover of assets,
T means sales,
A assets.

It expresses how effectively an enterprise can dispose of its assets regardless the sources of financing. The higher values of this ratio express more effective disposing of assets.

Inventory turnover is calculated from the following relation:

$$OZ\ =\ T\ /\ Z\qquad (4)$$

where OZ means inventory turnover,
T sales,
Z inventories.

It shows the extent to which an enterprise depends on its short-term assets. The higher the values the enterprise achieves, the faster the enterprise can monetize its current assets and is less dependent on it.

Profitability ratio

It includes: return on assets (ROA), return on equity (ROE), return on invested capital (ROI). Profitability ratios provide information regarding how effectively an enterprise can appreciate its assets with regard to equity or debt capital.

It will examine the following items from the balance sheet: total assets, equity, debt capital, costs for sold goods, items from profit and loss: HV for accounting period, sales.

The return on assets is calculated as follows:

$$ROA = EAT/A \tag{5}$$

where ROA means the profitability of assets,
EAT earnings,
A assets.

The calculation renders us find out how effectively the enterprise disposes of its total assets.

The return on equity is calculated as follows:

$$ROE = EAT/ E \tag{6}$$

where ROE means the return on equity,
EAT earnings,
E equity.

The calculation informs about to what extent the enterprise can appreciate its assets with the use of its own resources.

Return on investments is calculated in the following relation:

$$ROI = EBIT/ I \tag{7}$$

where ROI means return on investments,
EBIT earnings before interest and taxes,
I investments.

This ratio indicates the extent to which the enterprise can appreciate its invested capital.

Liquidity ratios

They include the following ratios: current ratio (BL), quick ratio (PL), cash position ratio (OL). Liquidity ratios will inform us about the extent to which the enterprise is able to fulfil its obligations from its short-term assets.

We will examine the following items from the balance sheet: inventories, short-term financial assets, current assets, short-term obligations.

Current ratio is calculated in the following relation:

$$BL = OA / KZ \tag{8}$$

where BL means current ratio,
OA current assets,
KZ short-term obligations.

It expresses the extent to which an enterprise is able to fulfil its short-term obligations from current assets.

Quick ratio is calculated in the following relation:

$$PO = (OA - Z) / KZ \tag{9}$$

where PO means quick ratio,
OA current assets,
Z inventories,
KZ short-term obligations.

The calculation is analogical to a current ratio calculation with the difference that inventories are deducted from current assets.

Cash position ratio is calculated in the following relation:

$$OL = FM / K \tag{10}$$

where OL means cash position ratio,
FM financial assets,
KZ short-term obligations.

The ratio provides information about the extent to which the company can fulfil its short-term obligations from its cash.

Further, we will use Altman Z-score, the ratio of Z factor should predict the bankruptcies of enterprises two years ahead, the further into the future we would like to predict, the lower the probability of bankruptcy. These models are very useful for an investor assessing the purchases of company shares or bonds. The models include important items such as assets, debts, profit, sales, obligations, etc. Table 1 shows that a financially stable enterprise is very likely to survive. The grey zone means that the probability to survive cannot be conclusively determined. The last result shows that an enterprise is immediately threatened by bankruptcy.

Altman Z-score

It is calculated as follows:

$$Z = 0{,}717 * X1 + 0{,}847 * X2 + 3{,}107 * X3 + 0{,}42 * X4 + 0{,}998 * X5 \tag{11}$$

where X1 is calculated as: net working capital / total assets,
X2 the profit of past years / total assts,
X3 EBIT / total assets,
X4 basic capital / total assts,
X5 sales / total assts,
EBITit means earnings before interest and taxes here

Table 1. The values of Altman X-score results.

Result	Assessment
> 2,99	financial stability
1,8 to 2,98	grey zone
> 1,7	bankrupting enterprise

Source: Authors.

4 Results

In the horizontal analysis of assets – relative, the year-on-year change of assets in percentage against the year of 2016 was observed. The horizontal analysis of assets – relative, it is shown in graph No.1. In 2017, fixed assets (DM) grew by 3% against the year of 2016 and subsequently dropped by 15% in 2018. Long-term intangible assets (DHN) were supposed to grow by 87% in 2017, it could have meant buying better software for the company, a decrease occurred in 2018, up to -38%, which is an unstable fall. Fixed assets (DM) grew by 3% in 2017 and subsequently dropped by 15% in 2018. The only item that failed to drop under the value equal to the year of 2016 are current assets (OM), they grew by 7% and by 26% in 2018. Inventories (ZAS) show a deep and unstable decrease, the fall in 2018 was by incredible 344% and it was equal to 0% in 2017%. The long-term debts (DPOH) are equal to 0%. Shor-term debts (KPOH) grew by 20% in 2017 against 2016 and subsequently dropped by 33%. Financial assets (FM) grew to the value of 87% in 2017 against 2016 and decreased by 1% in 2018. Comparing assets in total with manufacturing industry, the enterprise is below the usual values in the sector on average by CZK 1-1.5 million.

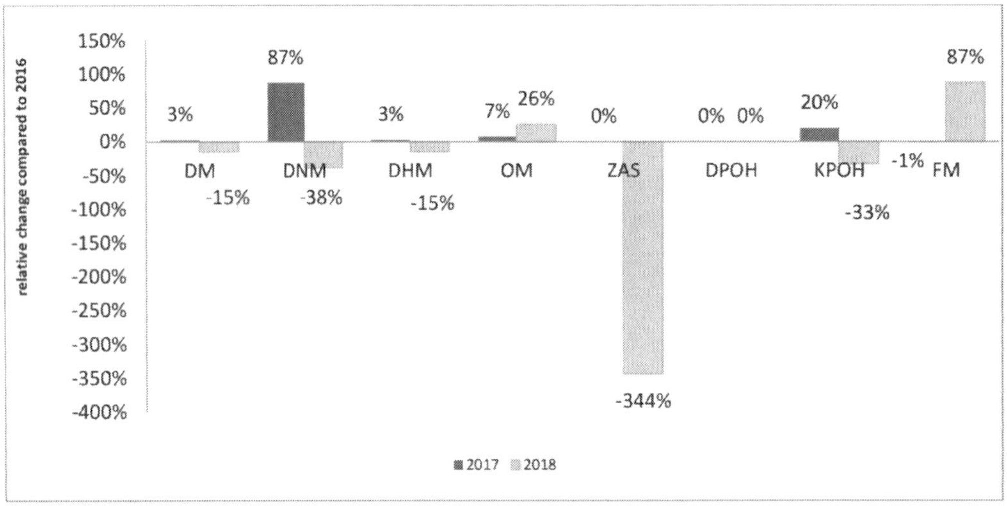

Graph 1. Horizontal analysis of assets – relative.
Source: Authors.

In the horizontal analysis of assets – absolute, we observe the growth or decrease of items in absolute units, in CZK thousands, against 2016. The horizontal analysis of assets is shown in graph No.2. Fixed assets (DM) decreased in 2017 and by CZK 55, 000, 000 in 2018. Long-term tangible assets (DNM) increased by CZK 47, 000, 000 in 2017 and by CZK 32, 000, 000 in 2018. Long-term tangible assets (DHM) grew by CZK 13, 517, 000 in 2017 and decreased by CZK 55, 359, 000 in 2018. Current assets (OM) increased by CZK 28, 000, 000 in 2017 and it also grew by CZK 164, 000, 000 in 2018. Inventories (ZAS) show 0 difference in 2017 against 2016, they decreased by CZK 2, 482. 000 in 2018. Long-term debts (DPOH) fail to demonstrate any difference in 2017 and 2018 against 2016. Short-term debts (KPOH) grew by CZK 20, 970, 000 in 2017 and decreased by CZK 5, 158, 000 in 2018. Financial assets (FM) fell by CZK 8, 000 in 2017 and grew by CZK 6, 897, 000 in 2018. Total assets increased by CZK 42, 323, 000 in 2017 and grew by CZK 116, 292, 000 in 2018.

 https://doi.org/10.1051/shsconf/202316001006

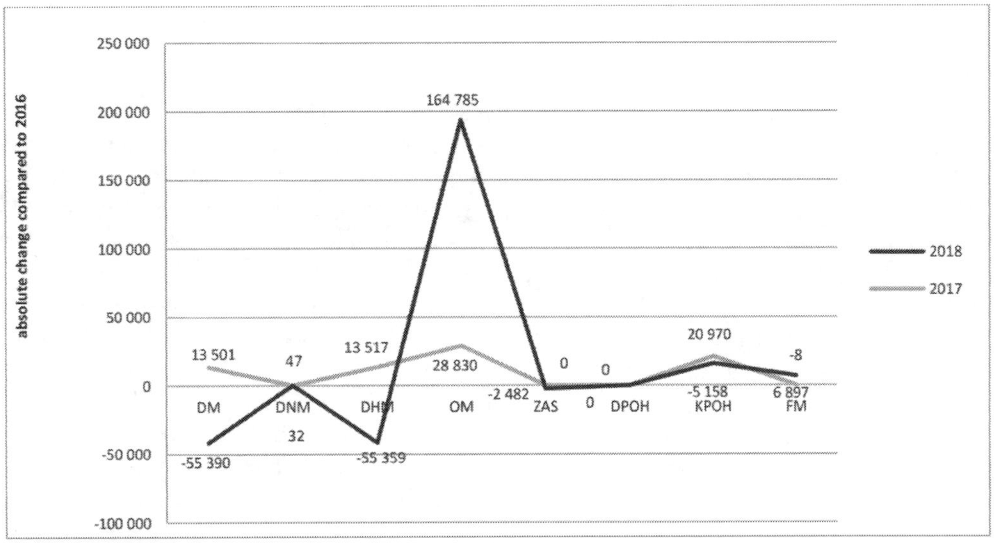

Graph 2. The horizontal analysis of assets – absolute.
Source: Authors.

The percentual expression of items from assets in total for individual years was calculated by vertical analysis. Vertical analysis is shown in graph No.3. Fixed assets (DM) accounted for 58% of total assets in 2016, 57% in 2017 and 46.15% in 2018. Long-term tangible assets (DHM) accounted for 58% of total assets in 2016, 57% in 2017, 46% in 2018. Long-term financial assets (DFM) accounted for 32% of total assets in 2016, 32% in 2017, and 46% in 2018. Current assets (OM) accounted for 42% of total assets in 2016, 43% in 2017 and 53.58% in 2018. The items of long-term intangible assets (DNM), inventories (ZAS) and long-term debts (DP) show the zero expression of total assets. Financial assets accounted for 32% of total assets in 2016, 32% in 2017 and 46% in 2018.

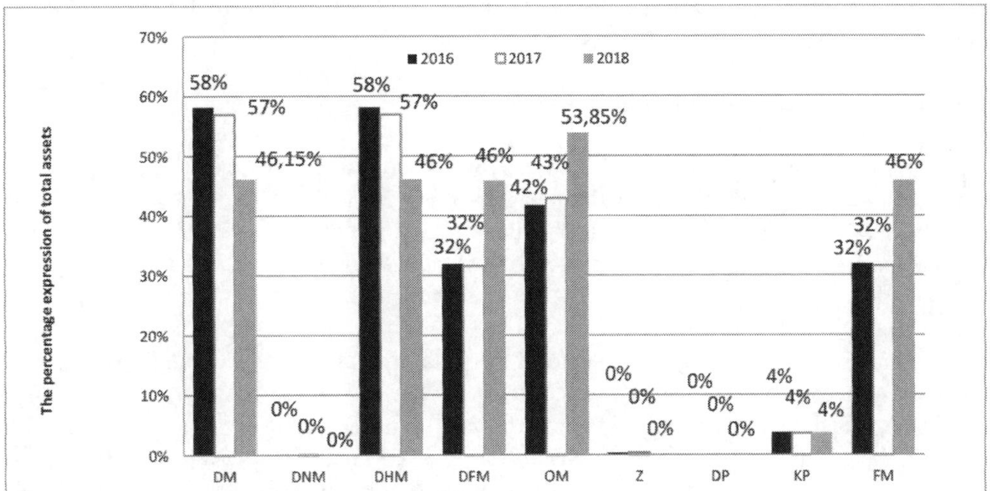

Graph 3. The vertical analysis of assets.
Source: Authors.

The analysis of ratios - assets

Activity ratios generally assess the way how the asset items of enterprise can transform into sales.

In 2016, the turnover of inventories changed into sales 180 x per annum, total assets turned into sales 0.66 x per annum and one item transformed into collection in 52 days. In 2017, the turnover of inventories changed into sales 208 x per annum, total assets turned into sales 0.71 x per annum and one item turned into collection in 56 days. In 2018, the turnover of inventories transformed into sales 986 x per annum, total assets turned into sales 0.71 per annum and 1 item changed into collection in 39 days.

The value of asset turnover should be at level 1 at least. The higher the value, the better the company uses its resources. If an enterprise has a law value of this ratio, it means that its asset structure is disproportional and is not used effectively.

In our case, the turnover of assets was 0.66 in 2016, 0.72 in 2017, 0.71 in 2018 and it was therefore lower than the time of inventory turnover, which was 2.00 in 2016, 1.73 in 2017 and 0.37 in 2018. It means that the enterprise is not able to efficiently dispose of assets and cannot efficiently use them and turn them into profit. The turnover of inventories is excellent in the enterprise. However, the time of debt collection is rather poor. The due date of fulfilling obligations is acceptable.

In comparison with manufacturing sector, the enterprise had a lower turnover of total assets, specifically, by 0.71 in 2016, 0.61 in 2017 and by 0.54 in 2018. The time of inventory turnover was higher in comparison with the sector, in the difference of 1.97 in 2016, 1.71 in 2017 and 0.34 in 2018. The time of debt collection was very high in the difference against the sector, by 50 days in 2016, by 55 days in 2017 and by 39 days in 2018.

The turnover of inventories is higher than in the sector, which is positive for the enterprise, it was higher in the difference of 90 days in 2016, 116 days in 2017 and 907 days in 2018. The due date of fulfilling obligations is also higher than in the sector, it was in the difference of 20.53 in 2016, 12.66 in 2017 and 18.35 in 2018.

We assess the enterprise to be worse off than manufacturing sector, in terms of the turnover of total assets, the time of inventory turnover, the due date of fulfilling obligations. In contrast, it is better off than the sector in terms of the turnover of inventories, the due date of fulfilling obligations. The enterprise has redundant assets located in current assets.

The analysis of ratios - profitability

The profitability of capital was at 5.23% in 2016 and subsequently rose to 8.23% in 2017 and to 9.17% in 2018. The profitability of total assets was 4.35% in 2016 and subsequently it rose to 6.6% in 2017 and to 7.48% in 2018. The profitability of equity was 5.02% in 2016, 7.45% in 2017 and 8.29% in 2018. The profitability of sales was 6.63% in 2016 and 9.22% and 10.46% in the following years. The last ratio is profit margin, which was 6.63% in 2016, 9.22% in 2017 and 10.46% in 2018.

In comparison to manufacturing sector the profitability of capital is overall lower than the value in the sector, specifically, by 8.96% in 2016, by 6.52% in 2017 and by 3.83% in 2018. The ROA of the enterprise requires lower figures than the sector, by 4% on average. The profitability of equity was lower than the sector by 3.02% in 2016, by 2.08% in 2017 and by 3.08% in 2018. The profitability of sales was below the sector level by 1.22% in 2016, but it showed higher values than the sector in the following years, specifically by 1.59% in 2017 and by 3.98% in 2018. The analysis of profitability renders us conclude that the enterprise is slightly below the level of its sector.

The analysis of ratios - liquidity

Current ratio accounts for 11.14 in 2016, 16.90 in 2017 and 16.64 in 2018.

Quick ratio is higher that it should be in enterprises, it is within the range of 1.2-1.6, the value was 11.04 in 2016, 16.76 in 2018 and 16.62 in 2018. Cash position ratio of the enterprise is right and corresponds to the values that should be, it was 0.03 in 2016, 0.04 in 2017 and 0.21 in 2018.

In comparison with the manufacturing sector, current ratio is very high, the difference was: by 9.46 in 2016, by 13.24 in 2017 and by 12.64 in 2018.

Quick ratio is also higher in comparison with the sector, it was by 10.02 in 2016, by 13.84 in 2017 and by 13.13 in 2018.

The recommendation is to decrease debts that are too big and fail to gain profit.

Altman Z-score

Altman Z-score appears in all three monitored years in the grey zone. The grey zone means that the probability of survival cannot be conclusively determined. The enterprise certainly has room for improvement.

In 2016, the enterprise shows the value of 1.58, which is a grey zone, in 2017, the value is 2.34, which is a grey zone as well, in 2018, the value is 2.44 and it is very close to being financially stable, the stability begins at the value of 2.99, however, the enterprise remains in the grey zone according to the figures in 2018. The recommendation is to decrease the volume of debts and to increase the volume of obligations.

Table 2. Comparison of ratio results.

Ratio	Enterprise			Sector			In a better position is the:
	2016	2017	2018	2016	2017	2018	sector/enterprise
The turnover of total assets	0,66	0,72	0,71	1,37	1,333	1,25	sector
The time of inventory turnover	2,00	1,73	0,37	0,03	0,03	0,02	sector
The turnover of inventories	180,14	207,73	985,85	90,00	92,00	79,00	enterprise
The time of debt collection	51,88	56,36	39,46	2,00	1,00	1,00	sector
The due data of obligations	20,54	12,67	18,36	0,01	0,01	0,01	enterprise
The profitability of capital	5,23%	8,23%	9,17%	14,19%	14,75%	13,00%	sector
The profitability of total assets	4,35%	6,66%	7,48%	8,21%	10,86%	11,48%	sector
The profitability of equity	5,02%	7,45%	8,29%	8,04%	9,53%	11,36%	sector
The profitability of sales	6,63%	9,22%	10,46%	7,85%	7,63%	6,48%	enterprise
Current ratio	11,14	16,90	14,64	1,68	3,66	2,00	enterprise
Quick ration	11,04	16,76	14,62	1,02	2,92	1,49	enterprise
Cash position ratio	0,03	0,04	0,21	0,24	0,75	0,54	enterprise

Source: Authors.

Table 2 shows the results of individual ratios. Even at first sight, it is clear that the company is not in a bad position in comparison with the results of the manufacturing sector. As for the turnover of assets, it had better results than the sector in the years of 2016, 2017 and 2018, therefore the enterprise shows worse results. At the time of inventory turnover, in the examined years, the sector achieved better results than the enterprise whose results were worse. As for the turnover of inventories, the enterprise is in a better position against the sector, the enterprise can turn the turnover of inventories into sales more times per annum than the sector. At the time of debt collection, the sector is better than the enterprise, the enterprise shows vary high values and is worse off than the sector. At the due date of obligations, the enterprise is better off, its figures are higher against the sector. In terms of the profitability of capital, the profitability of total assets and the profitability of equity, the sector is better off against the enterprise, however, the enterprise is below the level of the sector on average by 4%. Regarding the profitability of sales, the sector shows better values by mere 1.22% in the monitored year (2016), in further monitored years (2017, 2018) the enterprise is better. Concerning current ratio, the enterprise is better off against the sector. In quick ratio, the enterprise shows better values against the sector as well, the sector has very low values and is therefore worse off than the enterprise. In terms of cash position ratio, the enterprise is also better off than the sector.

5 Discussion of Results

On the basis of the obtained results, it was possible to answer the formulated research questions as follows:

VO1: How to determine redundant assets?

According to the ratios, as a unit, it may be concluded that the enterprise, in total, has quite positive results and an average asset structure. It failed to have quite optimal composition in every ratio and failed to always achieve as positive results as were in the sector, but it is a frequent phenomenon in practice. It may be found something that can be improved in every enterprise; it even applies to the best companies in the market. We arrived at the conclusion that the enterprise proves to have very good liquidity against manufacturing industry, but it should concentrate more on profitability and the turnovers of total assets and inventories. With the help of activity ratio, which was higher than in the sector, it can be concluded that the enterprise possesses redundant assets, specifically, the high volume of debts. [8] employed the method of asset efficiency analysis, in most cases they contained the systems of ratios, including the ratios of return on assets, the turnover of assets and the time of asset turnover. The method requires such interpretation of calculation results that takes into account the specifics of the sector where the enterprise does business. We do agree with this methodology, the results would otherwise be without an objective perspective and comparison. [21] used horizontal and vertical analysis of fixed assets. With the presence of further permanent factors, the part of assets increases, it may be concluded that the production capacity of enterprise expands. However, it is necessary remark that the increase in the asset part is not always an objective indicator of expanded production capacity. The change of tangible investment assets is often influenced by reconstruction, the division or merger of enterprise, changes in organisational structure and the system of management, the improvement of production processes and the forms of work organisation, the recovery of depreciated and obsolete assets, the level of specialisation in production and cooperation, the change of geographical location of business. Based on the analysis of Izos, limited liability company, we are unable to determine, whether the increase of asset structure was influenced by the innovation of asset structure or the increase of production, [21]. Tourism spending as well as innovations are characterized by great economic value, which emphasizes the need for their research from a macroeconomic perspective [23].

The influence of net profit and changed assets on the profitability of long-term tangible assets is crucially significant. There is a direct link between material fix and the profitability of assets and net profit. As the amount of net profit increases or decreases, the net profitability of assets correspondingly increases or decreases. Whereas, the relation between profitability and tangible investment assets is an inverse value. Decreasing the value of assets results in increasing the net profitability of tangible investment assets and vice versa – increasing the value of assets results in decreasing profitability. We arrived at this conclusion as well, there are direct links to property structure and the amount of net profit. Right at this point we observe the problem of Izos limited liability company, which is the size of debts that are tied to money the company can use somewhere else.

Redundant assets determined using the activity ratio appears to bet the volume of receivables that needs to be reduced.

VO2: What asset structure is the most appropriate for an enterprise?

It can be concluded from the results that the enterprise has quite good asset structure; however, it should change the volume of debts and increase the volume of inventories. [9] determined the financial performance of company by a number of various factors. It includes the policy of inventory management, which is designed in such a way that it determines the proportionate level and structure of inventories both in logistic and financial conditions. In economic practice, however, the strategy of inventory management significantly varies, e.g., Just in Time, Lean Management, Vendor Managed Inventory. It also differs in entrepreneurship sector, the size of company and industry. Studies based on the example of Polish food industry sectors showed there were statistically significant casual links concerning performance between inventory and financial days. In fact, the increase of these inventory cycles was significantly determining for ROA, rather than increasing days of other inventories. Briefly, this study found out that the management rationalization of inventories may be a significant power force for improving the financial performance of enterprises, therefore, it can contribute to creating values for owners. The size of inventories negatively influences the ratios of activity against manufacturing sector. We recommend Izos, limited liability company to increase the volume of inventories.

The enterprise has quite good asset structure, it was found out by using the ratios of financial analysis that the enterprise should adjust its current assets, specifically, to decrease the volume of debts and to increase the volume of inventories.

Within a case study, we recommend this adjustment of asset structure for Izos, limited liability company:

Fixed assets appear to be in an optimal proportion to the asset structure.

Current assets should be adjusted to generate the maximum possible profit, specifically, in terms of inventories and debts.

6 Conclusion

The objective of the paper was to determine redundant assets and the most appropriate asset structure for the enterprise. The objective was achieved by using the horizontal and vertical analysis of assets, the ratios of financial analysis and Altman Z-score. The results of the ratios of financial analysis were subsequently compared with manufacturing sector and it was searched where Izos, limited liability company would deviate from mean values and where the values would be in accordance. By vertical and horizontal analysis, it was discovered what asset structure the enterprise disposes of. The part of the data was used in calculating the ratios of financial analysis, which was essential for answering the research questions. Altman Z-score only proved whether the enterprise shows a future prediction for survival.

The effect of the results of activity ratio, profitability ratio and liquidity ratio enabled us to compare the results of manufacturing sector and determine the strengths and weaknesses of the enterprise. Due to this comparison, the deficiencies of the enterprise against the sector were determined. The recommended sizes of ratios are as follows: the turnover of total assets should equal 1, the time of inventory turnover should be the shortest possible, the turnover of inventories should be the highest possible, the time of debt collection should be the shortest possible, the due date of obligations should at least equal 2 or be higher.

The research limitation lies in the impossibility to determine the absolutely best asset structure but only the best possible structure on the basis on the comparison with the sector that sets the bar for the enterprises operating in the given sector.

Enterprises are thus recommended to monitor the activity ratios that directly influence the profitability of equity and its appreciation.

However, the right optimisation of asset structure can be even achieved by the minimisation of costs [5]. Creating the measures to reduce production costs is the main part of the cost management system in the enterprise; searching for reliable suppliers, improving products, better work organisation and sales, increasing work efficiency, maximising the involvement of additional income and accessible assets (renting sites not currently used by the enterprise results in obtaining additional income and, therefore, reducing the costs of maintenance of these premises or equipment), permanent controlling the costs of the enterprise, increasing the efficiency of internal cost management based on due time and high quality analysis; they are used for optimising operational costs in manufacturing enterprises.

References

1. T. Kovalchuk, A. Verhun, Improvement of the method of analysis of asset management efficiency. *Baltic Journal of Economic Studies,* **5**(5), 61-66 (2020)

2. M. M. Shygun, N. A. Ostapiuk, O. V. Zayachkivska, N. V. Goylo, The influence of the classification of non-current assets as holding for sales on the liquidity of the company's balance sheet. *Entrepreneurship and Sustainability Issues,* **8**(1), 430-441 (2020)

3. V. Spitsin, D. Vukovic, S. Anokhin, L. Spitsina, Company performance and optimal capital structure: evidence of transition economy (Russia). *Journal of Economic Studies*, **48**(2), 313-332 (2020)

4. K. Valaskova, M. Nagy, S. Zabojnik, G. Lazaroiu, Industry 4.0 Wireless Networks and Cyber-Physical Smart Manufacturing Systems as Accelerators of Value-Added Growth in Slovak Exports. *Mathematics*, **10**(14), 2452 (2022)

5. N. Bondarchuk, L. Vasiljeva, Impact of operating costs on economic phenomena and the possibility of their optimization at processors. *Baltic Journal of Economic Studies*, **4**(2), 40-46 (2018)

6. D. Tamulevičienė, Methodology of complex analysis of companies' profitability. *Entrepreneurship and Sustainability Issues*, **4**(1), 53-63 (2016)

7. A. Krause, Risk, capital requirements, and the asset structure of companies. *Managerial Finance,* **2**(9), 774-785 (2006)

8. N. O. Kovalchuk, M. G. Marych, L. V. Popova, A. O. Viknianska, Modeling of the optimal structure of meat processing enterprises. *Financial and credit activity: problems of theory and practice,* **4**(31), 270-278 (2019)

9. Z. Gołaś, Effect of inventory management on profitability: evidence from the Polish food industry. *Agricultural Economics*, **66**(5), 234-242 (2020)

10. R. Blazek, P. Durana, K. Valaskova, Creative Accounting as an Apparatus for Reporting Profits in Agribusiness. *Journal of Risk and Financial Management,* **13**(11), 261 (2020)

11. M. Munir, A. Kiviniemi, S. W. Jones, S. Finnegan, The business value of BIM for asset owners: a cross-case analysis. *Journal of Facilities Management,* **18**(5), 469-486 (2020)

12. N. L. Jacquet, Asset classes. *Journal of Political Economy*, 1-12 (2018)

13. A. E. Bernardo, A. Fabisiak, I. Welch, Asset redeployability, liquidation value, and endogenous capital structure heterogeneity. *Journal of Financial and Quantitative Analysis,* **55**(5), 1619-1656 (2020)

14. T. H. Van Hoang, C. Gurău, A. Lahiani, T. L. Seran, Do crises impact capital structure? A study of French micro-enterprises. *Small Business Economics,* **50**(1), 181-199 (2018)

15. O. Pakhnenko, O. Liuta, N. Pihul, Methodological approaches to assessment of the efficiency of business entities activity. *Business and Economic Horizons,* **14**(1), 143-151 (2018)

16. V. Jezkova, Z. Rowland, V. Machova, J. Hejda, The Intrinsic Value of an Enterprise Determined by Means of the FCFE Tool. *Sustainability,* **12**(21), 8868 (2020)

17. M. Vochozka, V. Machova, Enterprise Value Generators in the Building Industry. *Innovative Economic Symposium 2017 (IES2017): Strategic Partnership in International Trade*, **39** (2017)

18. D. Bauder, T. Bodnar, N. Parolya, W. Schmid, Bayesian mean–variance analysis: optimal portfolio selection under parameter uncertainty. *Quantitative Finance*, **21**(2), 221-242 (2021)

19. M. Hedvicakova, M., Kral, Benefits of KPIs for industry sector evaluation: The case study from the Czech Republic. *E+M Ekonomie a Management,* **22**(2), 97-113 (2019)

20. Ministry of industry and trade. *Financial analysis of the corporate sector for the year 2016* [online], Available at: https://www.mpo.cz/assets/cz/rozcestnik/analyticke-materialy-a-statistiky/analyticke-materialy/2017/5/FA4Q16.pdf (2017)

21. D. Tamulevičienė, J. Mackevičius, Methodology of complex analysis of tangible fixed assets. *Entrepreneurship and Sustainability Issues,* **7**(2), 1341-1352 (2019)

22. E. R. Lizarzaburu Bolaños, K. Burneo, L. Berggrun, Risk of insolvency and return of shares: empirical analysis of Altman's Z-Score in the Peruvian mining sector between 2008-2018. *Revista Universidad y Empresa*, **23**(40) (2021)

23. B. Gavurova, J. Belas, K. Valaskova, M. Rigelsky, V. Ivankova, Relations Between Infrastructure Innovations and Tourism Spending in Developed Countries: A Macroeconomic Perspective. *Technological and Economic Development of Economy,* **27**(5), 1072-1094 (2021)

The relationship between motives for using a Chatbot and satisfaction with Chatbot characteristics: An exploratory study

Tim Rieke[1] and *Helena* Martins[2, 3, 4, *]

[1]Faculty of Economics of Porto, University of Porto, Porto, Portugal
[2]Escola Superior de Ciências Económicas, Instituto Politécnico de Setúbal, Setúbal, Portugal
[3]CEOS.PP, ISCAP, P.Porto, Porto, Portugal
[4]NOVA SBE, Lisbon, Portugal

Abstract. Even though there is a growing number of studies focusing on Chatbots and artificial conversations, research lacks studies analyzing Chatbot characteristics and motives for using this technology. This displays a critical gap in the literature that the present study addresses. This work thus attempts to analyze the relationship between motives for using a Chatbot and satisfaction with Chatbot characteristics. Two questionnaires were developed, one to assess satisfaction with Chatbot characteristics, according to the Kano model and another to assessing motives for using Chatbots, based on previous qualitative research. Survey research was directed to the Portuguese population (N=258) and statistical analysis indicated that motives for using Chatbots do not seem to have a clear relationship with satisfaction with Chatbot characteristics. Furthermore, results suggest that equipping Chatbots with human-like characteristics, seems to be indifferent to Portuguese Millennials; instead, speed and accessibility of Chatbots seem to be valued, especially when using this technology for convenience purposes. A discussion on the possible implications for theory and practice on this topic is presented, and clues for future research are suggested.
Keywords: Chatbots, Artificial Intelligence, Natural Language Processing, Human-Computer Interaction

1 Introduction

Artificial Intelligence (AI) is on the rise, and it must be seen as an important IT pillar that will determine the competitiveness of companies shortly in what is commonly referred to as the 4th industrial revolution [1]. In general, three major technological trends now have come together and have led to the advancements in AI: big data, affordable high-performance computers and machine learning systems. Thus, AI is not just the future anymore it is part of our present [1].

Chatbots are dialogue systems that communicate with a user through natural language and a user interface [2]. Recently, these systems have experienced a new upswing [3].

* Corresponding author: helenagmartins@gmail.com

Researchers, entrepreneurs and individuals are using these digital helpers and see potential in improving the flow of information both inside and outside the company through the use of Chatbots [3]. Previously, customers who wanted to connect with a company either had to fill out forms or call hotlines with often long queues., generating communication that often can be one-sided, annoying and slow for the customer [4]. On the other hand, communication with friends and colleagues is increasingly taking place via messaging platforms [4]. The popularity of messaging and bot systems is steadily increasing. Since 2015, more people are using applications for communication in social networks. That is almost three billion people a day worldwide.

Consequently, a breakthrough into a new communication paradigm can be observed. Communication and interaction are increasingly controlled and determined by algorithms. Bots and messaging systems are vigorously debated and often have to serve as megatrends of the next few years. Ostensibly, it is all about new communication interfaces that bring efficiency and convenience benefits as the logical next evolutionary stage [5]. This is driven, in particular, by the advances in Artificial Intelligence, which make it possible to create learning algorithms and chatbots that can automate communication while being perceived "human-like" by the users [5].

This research aims to analyze the relationship between motives for using a chatbot and satisfaction with chatbot characteristics. Moreover, thus, to answer in what ways do motives for using a Chatbot affect the satisfaction with chatbot characteristics? If and how do the Chatbot characteristics that are valued by users vary with the motives for using them? Likewise, this interlinkage represents a gap found in the literature. Even though there is a growing number of research focusing on Chatbots and artificial conversations, the present literature review did not find of any studies analyzing the influence of various characteristics of chatbots on satisfaction.

2 Literature review

2.1 Chatbots

Chatbots, also called chatterbots, belong to the category of software agents as so-called interface agents or conversational agents. Chatbots allow humans to interact with the computer based on natural language, this can be text-based or voice-based [4, 5].
However, despite the diversity, all bots have one thing in common: they are text-based or voice-based systems that use Natural Language Processing to communicate with their users based on pre-defined rules or Artificial Intelligence. In the literature, there seems to be no general distinction or definition of Chatbot types [6, 5]. However, it can be stated that mainly two different types of Chatbots exist: rule-based-/Scripted-/Sequential bots and Intelligent (AI-based) bots.

The first type offers guided communication, using an existing set of pre-formulated rules and answers. Rule-based bots have an informative character. Since no artificial intelligence is used here, an open dialogue with them is not possible or at least only very limited. The programming effort is comparatively low [6]. Unlike the rule-based bots, the second type uses Artificial Intelligence techniques such as Machine Learning and NLP to understand enquiries. These bots learn the language much like a child and can create cross-references and recognise meaningful connections. They find answers to open questions and understand customer concerns, even without them having to be programmed exactly like that [6].

In the present work, the term "Chatbot" is used for all programs that enable natural language interaction. No matter if these are personified, whether the input works via keyboard, microphone or touchscreen. In addition to the language, additional elements of the communication are interpreted and integrated into the communication.

2.2 Chatbot characteristics

2.2.1 Emotion

A Chatbot uses natural language to interact with its opponent. It must be mentioned here, that the connection between language and emotion is still widely discussed today. While common sense suggests that language has nothing to do with emotion, it is evident that statements from other people affect our emotions. Likewise, humans use their words to describe their own emotions or those of others. Accordingly, it is believed that "this is the extent of the relationship between language and emotion" [7]. In this study *Happiness, Sadness* and *Empathy* are considered in this dimension.

2.2.2 Personality

Another trait that needs to be considered is the perception of users regarding the human body as a channel of communication, represented by an avatar in a Chatbot. There are several definitions of the term avatar. Bahosky and colleagues [8] describe it as "a pictorial representation of a human in a chat environment" (p. 8), whereas Loos [9] characterises avatars as "a representation of the user as an animated character in virtual worlds" (p. 17).

Studies and researches show that "human-like" behaviours associated with computer technology have an impact on the perception of the user. Due to language production, alternating conversation and reciprocal responding, users tend to personalise these technologies [10]. In this category: *Humour* and *Avatar* were considered.

2.2.3 Conversational abilities

A growing body of literature has examined artificial conversations regarding human-chatbot interaction. [11] for instance conducted a comparison between human-human online conversations and human-chatbot interactions and concluded that the interactions with a Chatbot had longer durations with shorter messages than the conversations between humans. Furthermore, their study points out that the communication with Chatbots "lacked much of the richness of vocabulary [...]and exhibited greater profanity" [11] (pp. 245). In this area, *Context Awareness* and *Common speech* were considered.

2.2.4 Efficiency

Chatbots are about new communication interfaces, which are bringing the next evolutionary step to efficiency and convenience benefits [12]. A significant advantage of using Chatbots is that companies can offer their services where most users are, such as in the messaging and social network apps [4]. The demands on the speed and competence of information have increased continuously in recent years [12]. Real-time service, as a customer's claim, has become a commonplace and indispensable part of effective service (Buttle & Maklan, 2015). As mentioned online and mobile communication has become a matter of course. Moreover, with that, additional expectations go along with the communication [12]. In this area, *Speed, Accessibility* and *Text vs. Voice* were surveyed.

2.3 Model

Given the literature review findings, this paper proposes the following model:

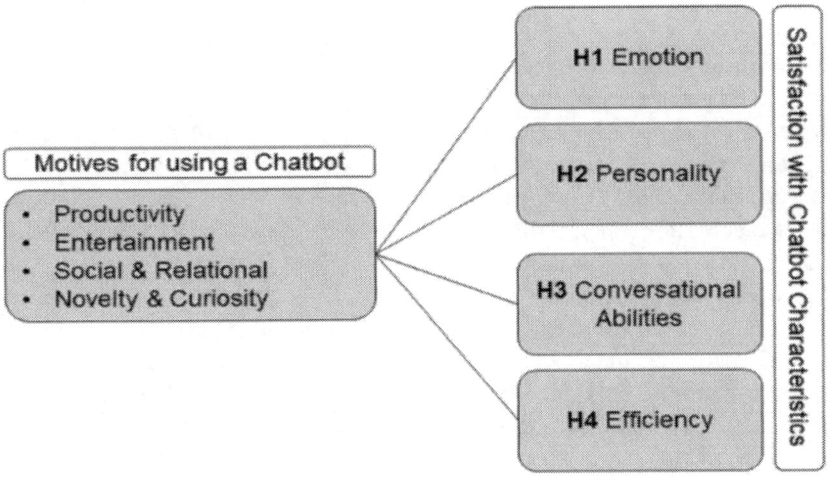

Fig. 1. Proposed model of the present study.

Source: Own processing.

Thus, the following hypotheses are derived from the model:
H1: The motives for using a chatbot influence satisfaction with the chatbot emotion.
H2: The motives for using a chatbot influence satisfaction with the chatbot personality.
H3: The motives for using a chatbot influence satisfaction with the chatbot conversational abilities.
H4: The motives for using a chatbot influence satisfaction with the chatbot efficiency.

3 Method

For the investigation of the hypotheses and research questions, a quantitative survey in the form of a questionnaire was chosen. To collect a larger sample in a timely and cost-effective manner, an online survey was conducted. The questionnaire was created with the service SurveyMonkey and made available online to the subjects.

To test and evaluate the questionnaire, a pre-test with talked discussion with a group of ten Portuguese Millennials was conducted before the main survey. This procedure aimed to find out if the questions are understandable for the subjects, if there are technical problems, useful comments and how long it takes to answer the questionnaire. As a result of the pre-test, some statements were adjusted for a better understanding of the items and four new questions concerning the participants' experiences made with a Chatbot were included. A convenience sample was gathered, since the survey was sent to university undergraduates, master-level students and alumni of the faculties of Economics and Engineering of the University of Porto, in Portugal and shared in social media (Facebook and LinkedIn platforms mainly). Only the Portuguese version of the survey was emailed and shared so that Portuguese could answer the survey with negligible linguistic barriers and in order to preselect and filter the target group.

Structure-wise the survey was composed of 52 items. The first six items are related to demographic information; four items concern the participants' experiences with a Chatbot 20 items are about the participants' motives of using a Chatbot and 20 items refer to the satisfaction with Chatbot characteristics. The survey also included two open questions, with the intention of allowing the participants comments, as well as email address, in case interest in receiving a summary of the survey results.

3.1 Assessing satisfaction

In this work we subscribe to the Kano model, which is a feature-oriented process and refers to product, service or interaction characteristics, judged by the user [13] This model is based on the two-factor theory and tries to determine the attributes that drive satisfaction [13]. The concerning attributes are classified according to their impact on satisfaction. There are three types of attributes that cause different levels of satisfaction [14]:

- *Must-be attributes:* the minimum criteria that must be fulfilled by a product, service or interaction. As a result, dissatisfaction is created in the event of non-fulfilment and neutral attitude upon fulfilment. Consequently, the satisfaction cannot be significantly improved if the performance level of a "Must-be" attribute continues to increase beyond the level expected by the customer or user [15].
- *One-dimensional attributes:* expected criteria and are thus explicitly requested. There is a linear relationship between the level of confirmation and satisfaction. This means that as the degree of fulfilment increases, satisfaction increases and vice versa. In highly competitive markets, only products or services that have high levels of performance attributes have a chance. As a result, performance attributes also play a prominent role in corporate communications. In many product areas, this attributes category serves above all to differentiate it from the competition [15].
- *Attractive attributes:* "Attractive" attributes are not expected and therefore not explicitly formulated and demanded. If a corporation succeeds in equipping its product with these attributes, this measure leads to positive disconfirmation and thus to a high level of satisfaction. Otherwise, there will be no sense of dissatisfaction when all the "Must-be" and "One-dimensional" attributes have the level of performance defined by the customer [15].

In addition, "Indifferent" and "Reverse" attributes are also of relevance in the Kano model. In the case of "Indifferent" attributes, there is no influence on the satisfaction, regardless of the fulfilment or non-fulfilment of the product characteristic. The "Reverse" attributes describe a category indicating that respondents do not desire the requested product feature and, in the presence, even leads to dissatisfaction [14].

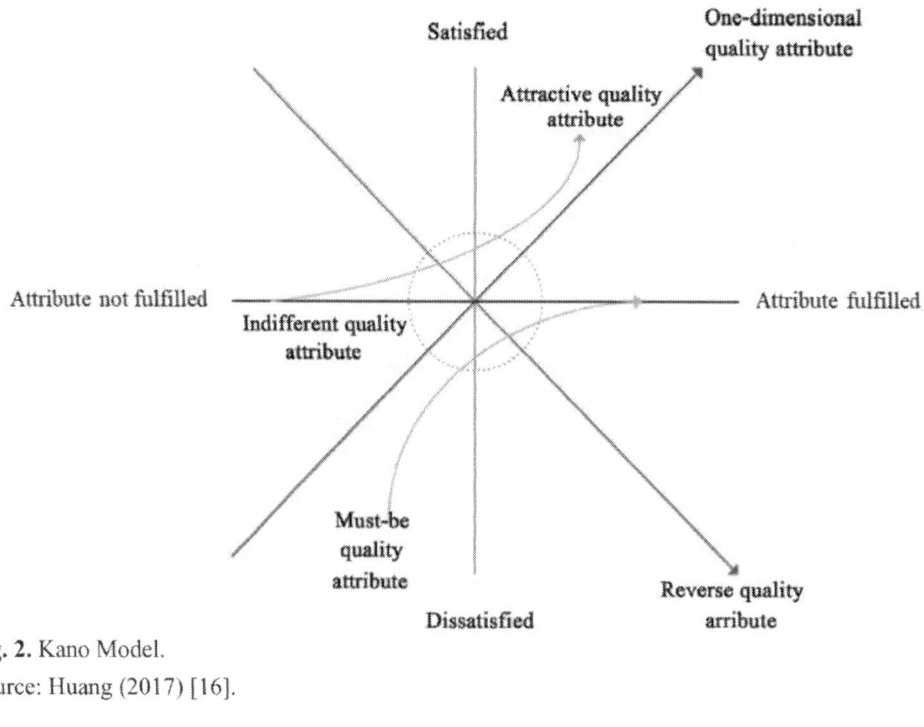

Fig. 2. Kano Model.

Source: Huang (2017) [16].

3.2. Sample

This study includes a sample of 258 valid responses, of which 126 (49%) identified as female and 131 (51%) identified as male. Most of the respondents were under 26 years old, totalizing over half of the sample (M = 25.36; SD = 4.37; Min = 18; Max = 38).

Most respondents report being familiar or very familiar with chatbots (65%), whereas 24% are a little familiar with this technology and 10% of respondents report not being familiar at all with chatbots. Regarding previous experience with chatbots, most respondents (54%) indicated a regular or very regular use of chatbots, 27% reported a rarely or sometimes using chatbots and 10% had no experience with chatbots. Most respondents 45% are curious or very curious about this technology, 47% are little or somewhat curious about chatbots and 8% are completely uninterested on the theme.

4 Results

The level of overall motives of the entire sample group to use Chatbots in general are Convenience (M=3,69), Exploring (M=2,96) and Social (M=1,87).
The evaluation displays that all analysed characteristics could be classified as indifferent-attributes and thus have no direct impact on satisfaction or dissatisfaction.

Table 1. KANO Model Assessment of CB characteristics.

Characteristic	A	O	M	I	R	Q	Dominant Category
Happiness	15	0	14	201	15	13	Indifferent
Empathy	16	3	14	201	14	10	Indifferent
Sadness	2	1	3	195	45	12	Indifferent
Humor	17	12	11	186	25	7	Indifferent
Avatar	11	3	21	206	11	6	Indifferent
Context Awareness	21	20	29	166	10	12	Indifferent
Common Speech	26	21	20	174	11	6	Indifferent
Speed	35	69	27	121	1	5	Indifferent
Accessibility	29	65	30	123	3	8	Indifferent
Text Based	15	11	18	194	9	11	Indifferent

A= Attractive quality attribute; O= One-dimensional quality attribute; M= Must-be quality attribute; I= Indifferent quality attribute; R= Reverse quality attribute; Q= Questionable.

Source: Own processing.

5 Discussion

This work has dealt with the research question in what ways motives for using a Chatbot may affect the satisfaction with Chatbot characteristics and if and how Chatbot characteristics that are valued by users vary with the motives for using them. All of the determined Chatbot characteristic were perceived as "Indifferent" attributes by the respondents, thus meaning that all of the original hypotheses should be rejected. This result may reflect that perhaps the elements we found in the literature might not be relevant for the subjects we surveyed.

Perhaps the literature around this topic has taken a somewhat "human-centric approach"" (in the sense that chatbots are conceptualized as ideally mimicking human affect, speech and appearance/avatar) but in fact chatbot users might not be looking for a humanoid bot anymore than people don't tend to expect their robot vacuum cleaners (e.g. Roomba) to look or behave like a human. Thus, we believe that these results might reflect a different conceptualization of what a chatbot is and should do in the users' perspective.

In line with these findings, subjects reported that "Emotion" or "Personality" traits were not relevant. Characteristics like happiness, sadness or empathy do not seem to cause an increase in satisfaction. Instead, the sample group valued Efficiency-related characteristics as important. This is confirmed through the analysis of the relationship between the motives and characteristics. All three motive dimensions display a positive effect on satisfaction in relation with Characteristics Accessibility and Speed. Another key finding is the absence of Entertainment as a motive to use Chatbots. While in the study by Brandtzaeg, and Følstad [17] the dimension "Entertainment" was assessed as an important motive, this dimension did not seem to make sense in the Portuguese population, rather, 3 new dimensions could be defined that can be used to capture motives for using a Chatbot, namely, Convenience, Exploring and Social, where the motives to use Chatbots for convenience purposes thereby displays the highest value.

With regards to the experience with Chatbots, the majority of the respondents (54%) reported having already had experienced interaction with a Chatbot, meaning nearly half the sample was referring to chatbots from an inexperienced or novice level perspective. The majority of participants reported only rarely using chatbots, despite indicating curiosity about the topic, which is confirmed in the literature [17]. Further confirming this idea, "exploring" could be identified as one of the three key dimensions regarding the motives to use a Chatbot. Contrary to other authors and developers who highlight the relevance of avatars, emotions, personality in chatbots and describe these characteristics as essential [6, 10], claiming that the goal must be to equip chatbots with human-like attributes to generate a successful interaction [12], it was interesting to realize that our sample does not seem to consciously value these aspects, but rather consider speed and accessibility as important features, delegating all other characteristics as indifferent in the Kano model.

6 Conclusions

Based on the research within this work it can be stated that Chatbot characteristics and motives for using this technology have been overlooked and only partly investigated in the literature which seems to focus more on the technical issues and less on perceptions of the user. There is still a gap in studies that address the implications of this technology and customer reactions to it, despite a growing interest in chatbot technology for experts, researchers, and companies [12], associated with a blooming expansion in the field of Artificial Intelligence in general and Natural Language Processing in particular.

Thus, this work aimed at exploring the relationship between Chatbot characteristics and motives for using them. Results show that characteristics such as efficiency and accessibility are of particular relevance to the selected target group. In general, subjects seemed to regard almost all of the other indicated characteristics as indifferent. One reason for this could be the prevalence of little knowledge of Chatbots and their preparation, namely that people might not realize that they are speaking or texting with a Chatbots or a virtual agent, something well documented in the literature.

It seems like the Chatbot technology serves mostly to obtain information quickly and efficiently and personality, emotions or an avatar have little relevance for the inquired subjects. These features contribute to increase costs associated with chatbots in terms of research and development as well as time to conclude projects; rather, our inquired users seem to be operating in a different paradigm where the features that really matter are speed and accessibility, contrary to a human-computer-interaction "humanoid" approach.

6.1 Limitations of the study and future research

The present study tried to create a framework for assessing chatbot consumer satisfaction and reasons for using chatbots. The methodology used proposed a novel approach to these elements by using the KANO model, but the way the scale was constructed did not allow for extensive statistical testing. Instead in future research we propose the scale be used with a 5-point Likert scale for the dimensions of the Kano model, in order for statistical procedures of scale validation to take place. The model should be further tested with a larger population size in order to acquire further validation and, ideally, further convergent and discriminant validity procedures should be undertaken. Another limitation was that a large portion of the sample did not have much experience with chatbots, which should be a variable that future studies should control for.

References

1. J. Kaplan. *Artificial intelligence*. Oxford: Oxford University Press (2016)
2. R. Dale. The return of the chatbots. Natural Language Engineering, **22**(5), 811-817 (2016)
3. E. Turban, J. Outland, D. King, J. Lee, T. Liang, & D. Turban, D. *Intelligent (Smart) E-Commerce. In Electronic Commerce 2018*. 249-283. Cham: Springer (2018)
4. O. Muldowney. Chatbots: An introduction and easy guide to making your own. Dublin, Ireland: Curses & Magic (2017)
5. S. Quarteroni. Natural language processing for industry. Informatik-Spektrum, **41**(2), 105-112. doi:10.1007/s00287-018-1094-1 (2018)
6. S. Janarthanam, *Hands-On Chatbots and Conversational UI Development: Build chatbots and voice user interfaces with Chatfuel, Dialogflow, Microsoft Bot Framework, Twilio, and Alexa Skills*. Packt Publishing (2017)
7. K. Lindquist, J. Maccormack & H. Shablack, The role of language in emotion: predictions from psychological constructionism. Frontiers in Psychology, **6**, 444. doi:10.3389/fpsyg.2015.00444 (2015)
8. R. Bahorsky, J. Graber & S. Mason, *Official Internet Dictionary: A Comprehensive Reference for Professionals Reviews*. Portland: ABS Consulting.
9. P. Loos (2003). Avatar. In Lexikon Electronic Business, Thomas Schildhauer, ed. Munich: Oldenbourg, 16–19 (1988)
10. C. Nass, & Y. Moon, Machines and mindlessness: Social responses to computers. Journal of Social Issues **56**: 81-103. doi:10.1111/0022-4537.00153 (2002)
11. J. Hill , W. Ford, & I. Farreras, Real conversations with artificial intelligence: A comparison between human–human online conversations and human–chatbot conversations. Computers in Human Behavior, **49**, 245-250. doi:10.1016/j.chb.2015.02.026 (2015)
12. P. Gentsch, Best Practices. In: *Künstliche Intelligenz für Sales, Marketing und Service*. Wiesbaden: Springer Gabler (2018)
13. Z. Li-Li, H. Lian-Feng, & S. Qin-Ying, Research on requirement for high-quality model of extreme programming. 2011 International Conference on Information Management,

Innovation Management and Industrial Engineering (ICIII), Shenzhen, IEEE, 26–27 Nov 2011 (2011)

14. N. Kano, N. Seraku, F. Takahashi, S. Tsuji. Attractive Quality and Must-be Quality, Hinshitsu: The Journal of the Japanese Society for Quality Control, 39-48 (1984)
15. A. Hussain, E. Mkpojiogu, F. Kamal, Eliciting user satisfying requirements for an e-health awareness system using kano model. In Xiaodong Z (ed) Recent advances in computer science. Proceedings of the 14th WSEAS International Conference on Computer and Computational Science (ACACOS'15), Kuala Lumpur. WSEAS Press (2015)
16. J. Huang, *Application of Kano Model in Requirements Analysis of Y Company's Consulting Project*. American Journal of Industrial and Business Management, 7, 910-918. doi:10.4236/ajibm.2017.77064 (2017)
17. P. Brandtzaeg, A. Følstad, Why People Use Chatbots. In Kompatsiaris I. et al. (eds) Internet Science. INSCI 2017. Lecture Notes in Computer Science, 10673. Cham: Springer (2017)

The relationship between unemployment and macroeconomic indicators

Eva Kalinová[1,*] and *Kateřina* Kroutlová[1]

[1]Institute of Technology and Business in České Budějovice, School of Expertness and Valuation Okruzni 517/10, 37001 Ceske Budejovice, Czech Republic

Abstract. The issue of unemployment set in the context of economic development is a crucial issue for ensuring economic growth of a country. The aim of the text is to analyze the relationship between unemployment and basic macroeconomic indicators and to confirm the direct relationship between these variables. The research method used was the analysis of the values of the general unemployment rate, gross domestic product, i.e. GDP, and gross value added, i.e. GVA, and their subsequent comparison. Other research methods include correlation analysis with Pearson's correlation coefficient, which confirmed a high degree of correlation between the general unemployment rate and the values of GDP and GVA. The research confirmed the existence of a correlation between the general unemployment rate and both GDP and GVA. GDP and GVA in the Moravian-Silesian Region were increasing in the sample years 2013-2019, while the general unemployment rate was decreasing. The analysis of the values shows that the values of GDP and GVA were inversely proportional to the values of the general unemployment rate. A limitation of the research was that economic development is not only influenced by the unemployment rate. As a result, it was not possible to establish a precise value that would define the extent of the impact of unemployment on the economic development of the region. The contribution of the paper is the confirmation of the existence of the correlation of unemployment with the mentioned development indicators, and the validity of Okun's Law also on the territory of the Moravian-Silesian Region.
Keywords: unemployment, GDP, GVA, Okun's law, Moravskoslezský region

1 Introduction

The topic of unemployment analysis in the context of the development of the Moravian-Silesian Region was chosen deliberately. Unemployment is a relatively large and long-term social problem in this region. However, according to [1], unemployment is not only a social risk, but also affects the national economy, which does not use the potential of the labour force due to the lack of jobs. The relationship between economic growth and unemployment shows that there is a high degree of correlation between the rate of economic growth and the decline in the unemployment rate [2]. The relationship between economic growth and

* Corresponding author: kalinova@mail.vstecb.cz

unemployment has been experimentally studied in the economic literature based on what is known as Okun's law, which shows that there is an inversely proportional relationship between a change in the growth rate (GDP) and a change in the unemployment rate [3].

The main research question is the impact of the general unemployment rate on the gross domestic product of the Moravian-Silesian Region. In this research part we try to confirm the validity of Okun's law in the selected area. The second research question is the effect of unemployment rate on gross value added of the same region. In this case, we try to prove the validity of Okun's law using a different economic indicator than gross domestic product. Both GDP, Gross Domestic Product, and GVA, Gross Value Added, are basic macroeconomic indicators used to determine the performance and the economy.

2 Literature research

The topic of unemployment is quite a debated topic and many professional and popular publications have been written about it. Unemployment is a serious social and economic phenomenon that is often perceived as causing a huge burden on society [4]. The impact and reflection of the Great Depression, in which unemployment was a big concept, was the focus of [5]. According to the authors, unemployment will always accompany us and in the future, we will have to deal with unemployment and its impact. The unemployment rate is one of the key indicators of the labour market, which contains information about the activity of the labour market and the overall state of the economy [6]. [7] focused on the sensitivity of unemployment to output as a function of time. According to his research, the latter was increasing, but this was in the context of greater use of fixed-term contracts [7].

The unemployment rate is the main macroeconomic indicator used in forecasting losses. The COVID-19 pandemic has caused an unprecedented level of volatility in labor market variables, leading to new challenges in using the unemployment rate in risk modeling [8].

While economists and academics argue convincingly that a certain natural rate of unemployment cannot be erased, increased unemployment imposes high costs on individuals, society, and the country [9]. Worse, most of the costs are of the deadweight loss type, where there are no offsetting gains with costs that everyone must bear. Depending on how it is measured, the unemployment rate can be interpreted [10].

It is true that: unemployment has costs to society that are more than just financial, unemployed people not only lose income but also face physical and mental health problems and there are social costs of high unemployment [11].

In advanced economies, macroeconomic indicators of unemployment and price indices tend to predict the sustainability of the relationship between them [12].

The topic of the impact of unemployment on mental health was also addressed by [13], whose work focused on different stages of unemployment. "Various fluctuations in mental health during periods of unemployment have been noted. We argue that there is a coping process in this situation" [13, 14]. Insurance can also stave off the fear of unemployment. Its influence has been the focus of [15], who describe the positive effects of unemployment insurance, which has been shown to reduce fears of future unemployment [15].

Macroeconomic variables respond to shocks in the perception of three dimensions of uncertainty (economic, inflation and employment) [16].

The cost of unemployment for individuals is not hard to imagine. When a person loses their job, it often has an immediate impact on their standard of living. Even those who are entitled to unemployment and other social benefits find state assistance insufficient, as these benefits often replace only part of their regular income. As a result, unemployed people consume much less than usual. Prolonged unemployment can lead to an erosion of skills and work habits, depriving the economy of useful talent that could be used in the future. The experience of unemployment (whether direct or indirect) can also change the way workers plan for the future - long-term unemployment can lead to greater scepticism and pessimism. In a similar vein, the absence of income generated by unemployment can lead families to deny their children educational opportunities, depriving the economy of a quality potential workforce [3]. [17] quantified the costs that are spent annually from the state budget in case of unemployment of young graduates. Analysis was performed using the counterfactual Propensity Score Matching method, followed by measures of employability and job sustainability at the end of the intervention.

The minimum wage is influenced by state intervention, which intervenes in the market mechanism and represents an important indicator when assessing the impacts in the wage area. If the average wage indicator is an endogenous variable, this variable has a positive as well as a negative impact on the remaining analyzed indicators [18].

The gender disparities in unemployment are discussed in [19], who finds that the main cause of these disparities is differences in the composition of the industry itself.

Its actual impact is looked at by [20], who, among other things, explain the basic principle as "the dynamics of output and unemployment that are relevant to markets, policymakers, and the average economic agent." [20] the very dominance of the output growth-unemployment relationship was verified in their paper by [21], who used Okun's Law to analyze these factors in the Netherlands.

Accurately identifying economic recessions is a macroeconomic challenge. The Sahm recession indicator is one of the most reliable early detectors, relying on changes in unemployment subject to labor force misclassification errors [22].

Due to the COVID-19 pandemic, companies, not only in cultural management, and their employees had to deal with a number of changes and people faced new challenges, not only in the workplace, but also in the area of unemployment [23]. The COVID-19 pandemic has affected every aspect of consumer behavior – their spending, investments, and financial reserves, as well as their financial and social well-being [24].

3 Data and methods

As the first data source, will be used data on the general unemployment rate in the Moravian-Silesian Region. The data source is the publicly accessible database of the Czech Statistical Office, which can be found on its website. It will be focused on examining the data for the years 2013-2019, where we will examine the general unemployment rate expressed as a percentage.

The second set of data that will be used as an indicator of the development of the Moravian-Silesian Region will be GDP and GVA, i.e., gross domestic product and gross value added. These are key indicators of economic development. The source of these data will also be the publicly accessible database of the Czech Statistical Office, which, as aforementioned, can be found on the CSO website. In this case will be also focused on data obtained from 2013-2019.

In this thesis, it will deliberately not be used the years 2020-2021, as this is the period when the economy slowed down to a complete halt due to the coronavirus epidemic. Therefore, the obtained data would not be conclusive.

The method of data collection will be compiled in this study of statistical data from the Czech Statistical Office. Subsequently, it will be made a comparison of the data. In this type of data comparison, it will be not necessary for the data to be expressed in the same unit, as it will only be important to express the growth or decline of a given value. The aim of the data comparison will be confirmed that unemployment has a major impact on the development of the region. The results will be displayed in tables.

It will be assumed that unemployment in the Moravian-Silesian Region will have a negative effect on local GDP and GVA. If the general unemployment rate falls, both values will rise. Therefore, these two indicators will be inversely proportional to the unemployment rate.

The next analysis will be a correlation analysis between the general unemployment rate and the evolution of GDP and GVA over the period 2013 to 2019. This analysis will be used the Pearson correlation coefficient. It will be denoted this coefficient by r. If the correlation (r) = 0, then there is no linear correlation between the correlation variables.

For a linear correlation to exist, r < 0, r > 0, r < 1 and r > -1 must hold.

If: r < 0, it is a negative correlation and if r > 0, it is a positive correlation.

In general, the coefficient r has to be in the interval [1;1] and the following formula will be used:

$$r = \frac{\sum_i (x_i - \bar{x})(y_i - \bar{y})}{\sqrt{\sum_i (x_i - \overline{x})^2}\sqrt{\sum_i (y_i - \overline{y})^2}} \quad (1)$$

The analysis will be processed in Microsoft Excel using data tools with the "Correl" function. The correlation coefficient will be divided into four groups of low correlation, medium correlation, high correlation and pure correlation.

Low correlation	r = 0.001 to 0.300 and -0.001 to -0.300
Mean correlation	r = from 0.300 to 0.750 and from -0.300 to -0.750
High correlation	r = 0.750 to 0.990 and from -0.750 to -0.990
Net Correlation	r = [1] and [-1]

4 Results

Based on the study of the unemployment rate in the Moravian-Silesian Region, was found the highest value in 2013. According to the Czech Statistical Office, in that year the unemployment rate was 9.9%. Then in 2014 the unemployment rate dropped to 8.6% and the downward trend continued. In 2015, the unemployment rate was 8.1%, in 2016 it fell to 6.9%, in the following year 2017 it continued to fall to 4.7% and in 2018 the unemployment rate was 3.7%. The lowest unemployment rate in the Moravian-Silesian Region was measured in 2019. In that year, the value fell to 3.6%. The development of the unemployment rate is shown in Table 1. Comparing the data from each year, we can notice that the general unemployment rate in the Moravian-Silesian Region has been decreasing in the long term.

Table 1. General unemployment rate 2013-2019.

General unemployment rate (%)							
Year	2013	2014	2015	2016	2017	2018	2019
Moravian-Silesian Region	9,9	8,6	8,1	6,9	4,7	3,7	3,6

Source: own processing based on data from the Czech Statistical Office.

The next value examined was GDP, i.e. gross domestic product. In 2013, GDP was 390,709 million. CZK. Then in 2014 it rose to CZK 416 291 million. CZK and it increased. In 2015, it reached CZK 435 540 million. In 2016 it reached CZK 450 316 million. In 2017, it reached CZK 471 366 million. CZK 3,366 million. In 2018, the GDP of the Moravian-Silesian Region exceeded CZK 500 000 million. The value in this year was 500 421 mil. CZK. In 2019, GDP reached 509 646 mil. CZK and thus the highest value of the examined years. The development of the value of GDP is clearly shown in Table 2. After comparing the values from these years, it reported that the value of GDP in the Moravian-Silesian Region was steadily increasing.

Table 2. Gross domestic product 2013-2019.

GDP (million CZK)							
Year	2013	2014	2015	2016	2017	2018	2019
Moravian-Silesian Region	390 709	416 291	435 540	450 316	471 366	500 421	509 646

Source: own processing based on data from the Czech Statistical Office.

The last value dealt with was GVA, gross value added. It amounted to 350 175 million in 2013. CZK. In 2014 it rose to CZK 376 519 million. It continued to grow. In 2015 it was CZK 392 206 million. In 2016, it reached CZK 405,053 million. CZK 423,579 million in 2017 and CZK 423,059 million in 2014. CZK 450 964 million in 2018. CZK. It then reached its peak in the last year under review, reaching CZK 460 479 million. CZK 460 479 million. The value of GVA as well as the value of GDP has been increasing every year. The GVA values in each year were shown in Table 3.

Table 3. Gross value added 2013-2019.

GVA (million CZK)							
	2013	2014	2015	2016	2017	2018	2019
Moravian-Silesian Region	350 175	376 519	392 206	405 053	423 579	450 964	460 479

Source: own processing based on data from the Czech Statistical Office.

Table 4. Correlation analysis between GDP (million CZK) and the Moravian-Silesian Region between 2013-2019.

	Year (2013-2019)	GDP (mill. CZK)
Year (2013-2019)	1	
GDP (million CZK)	0,995704587	1

Source. own processing based on data from the Czech Statistical Office.

In Table 4 was find the value of the correlation coefficient between GDP (in million CZK) and the period 2013 to 2019. The value of the correlation coefficient is 0.995704587. This value was close to the net correlation.

Table 5. Correlation analysis between GVA (million CZK) and the Moravian-Silesian Region between 2013-2019.

	Year (2013-2019)	GVA (million CZK)
Year (2013-2019)	1	
GVA (million CZK)	0,994570331	1

Source. own processing based on data from the Czech Statistical Office.

In Table 5 was find the value of the correlation coefficient between GVA (in million CZK) and the period 2013 to 2019. The value of the correlation coefficient was 0.994570331. This value was close to the next correlation.

The first research question was the impact of unemployment on the development of the Moravian-Silesian Region, where the indicator of the unemployment rate was the general unemployment rate and the indicator of development was the gross domestic product. These values would be inversely proportional. From the measured values and from the analysis it was obvious that the general unemployment rate has been steadily decreasing from 2013 to 2019, while GDP has been increasing. This forecast was therefore confirmed.

The second research question was also the impact of unemployment on the development of the region, but this time the indicator of development was gross value added. As well as GDP, GVA is also growing between 2013 and 2019. Thus, the assumption was confirmed in both cases. Both GDP and GVA are inversely proportional to the general unemployment rate. The higher their value, the lower the share of unemployed persons in the total labour force.

This fact points to a correlation between unemployment and the economic development of the region, which was further examined in Tables 4 and 5. The correlation coefficient was examined between the general unemployment rate and the GDP in the period 2013-2019, and between again the general unemployment rate and the GVA in the period 2013-2019. The resulting value of both coefficients exceeded 0.99, which was close to a net correlation with the chosen methodology and thus the association was confirmed.

The research confirmed the results of the literature and proved the validity of Okun's law. The validity was demonstrated in the relationship of unemployment with gross domestic product as well as in the relationship of unemployment with gross value added.

5 Conclusions

The research achieved its aim by answering both research questions. The first one was the impact of unemployment on the development of the Moravian-Silesian Region, where the indicator of development was the gross domestic product. The second one was the impact of unemployment on the development of the Moravian-Silesian Region, when gross value added was used as the development indicator. The assumption was an inverse relationship between the general unemployment rate and GDP and GVA.

The assumption was confirmed by research. While the unemployment rate declined between 2013 and 2019, gross domestic product and gross value added increased.

Evidence that the objective of the thesis has been met the numerical results that have demonstrated the validity of Okun's law of correlation between the unemployment rate and gross domestic product. It was shown that Okun's law can be applied only to a selected area, in this case the Moravian-Silesian region, and cannot be applied only to the national economy. Furthermore, it was proved that this law can be applied to research with other indicators of economic development than gross domestic product.

A limitation of the research was factors that affecting the economy of the region itself. Thus, the economic growth of the Moravian-Silesian Region was not logically influenced only by the unemployment rate. Due to this, the research method used cannot provide a precise value that would define the extent of the impact of unemployment on GDP and GVA.

The research can serve mainly as a basis for further and deeper investigation of the relationship between unemployment and GDP and GVA in the Moravian-Silesian Region.

Acknowledgements

This research was supported by the project, which has received funding from the Institute of Technology and Business. Grant Number IVSUZO001 "Benchmark of the world economy in the context of the principles of the circular economy and determining the value of the company". This paper was also supported from the project of Specific University Research, reg. number 05SVV22 – Scientific School and its connection with the experts from practice in the research activities of the institution.

References

1. H. Kolibová, A. Kubicová, *Labour market and employment policy: distance learning support.* Karviná: Silesian University in Opava, Faculty of Business Administration in Karviná (2005)
2. V. Machová, J. Horák, GDP as a risk factor of Czech enterprises. *Proceedings of the 13th International Management Conference: Management Strategies for High Performance (IMC2019): International Management Conference*, 785-798 (2019)
3. M. Verbeek, *A guide to modern econometrics*. 4th ed. Chichester: John Wiley & Sons, XV, 497 (2012)
4. P. Kuchař, L. Vaška, *Regional Aspects of Unemployment in the Czech Republic and Slovakia.* Jan Evangelista Purkyně University (2013)
5. N. Petrovsky-Nadeau, L. Zhang, Unemployment crises. *Journal of Monetary Economics,* **117**, 335-353 (2021)
6. G. A. Monusová, Perception of Unemployment by Europeans. *World Economy and International Relations,* **64**(4), 84-95 (2020)
7. K. Bartošík, Temporary contracts and Okun's law in Poland. *Equilibrium. Quarterly Journal of Economics and Economic Policy*, **15**(4), 735-760 (2020)
8. Z. Liu, H. Liang, Permanent Layoff And Consumer Credit Card Loss Forecasting. Managerial Finance (2022)
9. J. Vrbka, G. A. P. Lutan, Evaluation of the effects of intervention programs for unemployed: An overview of methods. *Ekonomicko-Manazerske Spektrum*, **13**(2), 56-67 (2019)
10. S. Hašková, P. Volf, V. Machová, Economic convergence of Czech regions in terms of GDP and unemployment rate in response to FDI flows: Do businesses and regions flourish? *Ad Alta: Journal of Interdisciplinary Research*, **9**(1), 326-329 (2019)
11. R. Holman, *Macroeconomics: an intermediate course.* 2nd edition: C. H. Beck, XIV, 424 (2010)
12. Z. Košťálová, E. Horvátová, Š. Lyócsa, P. Gernát, New credit drivers: Results from a small open economy. *Eastern European Economics*, **60**(1), 79-112 (2022)
13. S. Thill, C. Houssemand, A. Pignault, Unemployment Normalization: Its Effect on Mental Health During Various Stages of Unemployment. *Psychological Reports*, **122**(5), 1600-1617 (2019)
14. G. Malisauskaite, O. Nizalova, D. Xanthopoulou, Unemployment and well-being of Europeans across the life cycle: The role of countries' macroeconomic situation. *Social Indicators Research*, **162**(3), 1387-1412 (2022)
15. J. Ng, T. Ranasinghe, G. Shi, H. Yang, Unemployment Insurance Benefits And Income Smoothing. *Journal Of Accounting And Public Policy*, **38**(1), 15-30 (2019)

16. O. Claveria, Uncertainty indicators based on expectations of business and consumer surveys. *Empirica*, **48**(2), 483-505 (2021)
17. L. Švábová, M. Durica, T. Kliestik, Modelling the Costs of Unemployment for Young Graduates in Slovakia: A Counterfactual Approach. *Politicka Ekonomie*, **67**(5), 552-566 (2019)
18. L. Meixnerová, M. Krajňák, Macroeconomic time series affecting the minimum and average wages of V4 countries. *E+M Ekonomie a Management*, **23**(4), 4-22 (2020)
19. S. Albanesi, A. Şahin, The gender unemployment gap. *Review of Economic Dynamics*, **30**, 47-67 (2018)
20. L. Donayre, On the behavior of Okun's law across business cycles. *Economic Modelling,* **112** (2022)
21. J. P. Elhorst, S. Emili, A spatial econometric multivariate model of Okun's law. *Regional Science and Urban Economics,* **93** (2022)
22. J. Sun, S. Feng, Y. Hu, Misclassification Errors In Labor Force Statuses And The Early Identification Of Economic Recessions. *Journal Of Asian Economics*, **75** (2021)
23. P. Horváthová, K. Mokrá, H. Štverková, Remote Work and Other Effects of the COVID-19 Pandemic in Enterprises and Cultural Management. *Cultural Management: Science and Education,* **6**(1), 129-153 (2022)
24. K. Valášková, P. Durana, P. Adamko, Changes in Consumers' Purchase Patterns as a Consequence of the COVID-19 Pandemic. *Mathematics,* **9**(5), 1788 (2021)

Research on the development path of China's cross-border e-commerce industry under the COVID epidemic

Jiří Kučera[1,*] and *Yan* Zhang[1]

[1]Institute of Technology and Business in České Budějovice, School of Expertness and Valuation, Okruzni 517/10, 37001 Ceske Budejovice, Czech Republic

Abstract. The COVID-19 epidemic in 2020 has swept the world, bringing great challenges to Chinese cross-border e-commerce industry, as well as opportunities for development. According to my country's economic survey in the first and second quarters of 2020, the COVID-19 epidemic has caused a sharp drop in the transaction volume of Chinese foreign trade industry, while the cross-border e-commerce industry has grown against the trend. Starting from the characteristics of Chinese cross-border e-commerce industry, this article analyzes the impact of the COVID-19 epidemic on Chinase cross-border e-commerce industry, mainly in the obstruction of logistics and the impact on the global supply chain. Taking Ali International Station as a case analysis, through the formulation of support policies for sellers, investment in digital trade and other measures to help e-commerce sellers to tide over the difficulties, and then put forward suggestions on path selection for Chinese cross-border e-commerce industry to cope with the COVID-19 epidemic.
Keywords: COVID-19; cross-border e-commerce; Ali International Station

1 Introduction

The scale of China's foreign trade has continued to expand, the number of export commodities has also continued to increase, and the number of export cross-border e-commerce (EC) industries has also grown by leaps and bounds. According to e-commerce data released by the Ministry of Commerce, the scale of China's export cross-border e-commerce transactions in 2019 reached 8.03 trillion yuan. An increase of 13.09% compared with 2018, China's cross-border e-commerce industry continues to rise, cross-border e-commerce has higher and higher requirements for capital, technology and services, and the industry threshold and operational difficulty also increase accordingly.

More traditional enterprises and brands are beginning to move towards the center of the stage, the cross-border e-commerce industry began to develop in the direction of branding and diversification. "Therefore, if the EC market develops in a country, the negative impact of COVID-19 on import demand will decrease in that country." [1] The important parallel between import and exported were heavily impacted due to the pandemic of COVID-19.

* Corresponding author: kuceraj@mail.vstecb.cz

"Although China's exports continued to decline, cross-border electronic commerce's exports went against the trend." [2] Nacar and Ozdemir [3] n their study focused on relation between of e-commerce marketplace website and the effect of the COVID-19. The COVID-19 pandemic heavily impacted the e-commerce market and changed the near future in shopping. "In the context of the global epidemic and economic structural reform and transformation, China has proposed a new economic pattern of "dual circulation" development, and this measure has greatly promoted the development of China's cross-border e-commerce industry" [4]. China and other countries all around the world were forced to change their market structures. But the pandemic didn't affect only the companies. "The Corona Virus Disease (COVID)-19 pandemic has disrupted the business and industry landscape and changed consumers' behavior." [5]. The customers behaviour was drastically changed due to the pandemic of COVID-19. "Overall, consumers recognized the importance of e-commerce immediately after the pandemic, with no subsequent significant decline. Second, there may be a positive relationship between the time spent at home and the importance of e-commerce One reason is that an increase in stay-at-home duration decreases the opportunity to shop at retail stores." [6]. Stronger; On the other hand, policy dividends from the cross-border e-commerce comprehensive pilot zone. "The strict traffic control and personal isolation measures were effective in hindering the spread of COVID-19 in China" [7]. Economic sanctions are among the most powerful instruments of international policy [8].

"Limitations on the mobility of workers, variations in consumer demands, the shutdown of food manufacturing industries, limited food trading regulations, and financial stress throughout the food supply chain are all results of the COVID-19 pandemic." [9] It has a greater impact on the expectations and confidence of market players. Zhiyong [10] said that when the epidemic affects the centralized office model, the gradually mature online office has become the first choice of many enterprises, and remote office has ushered in a new "inflection point" of development, which will promote the rapid development of the industry. Nowadays, telecommuting is on the rise, and the awareness of enterprises and employees is also increasing. The way of working is only a form, and more importantly, it is to improve work efficiency. The COVID-19 pandemic, called a global health crisis, was an unpredictable risk of global proportions that paralyzed the entire world [11].

In 2020, affected by the epidemic, the risk of supply chain and logistics costs is relatively high, the risk of depending? Default on customer funds is rising, many small and medium-sized enterprises are facing bankruptcy, and the overall purchasing and consumption capacity will decline. "For e-commerce enterprises, logistics cost is an important factor that affects the operation effect and consumer experience." [12] Therefore, by analyzing the development of cross-border e-commerce retail enterprises, as well as the import and export data of cross-border e-commerce from May 2019 to December 2020, we will analyze the problems existing in the cross-border e-commerce industry and propose solutions.

It is necessary to prepare internal corporate environment in context to digitization, when trades enter to unknown foreign markets [13].

2 Characteristics of China's cross-border e-commerce industry

From 2015 to 2019, the total value of China's cross-border e-commerce import and export retail sales continued to expand. In 2019, the total value of China's cross-border e-commerce retail imports and exports continued to rise, reaching 186.21 billion yuan, an increase of 38.3% over the previous year. Affected by the COVID-19 epidemic, the cross-border e-commerce industry has been greatly impacted in 2020, and the total value of China's cross-border e-commerce retail imports and exports has decreased compared with the previous year.

From 2014 to 2019, the scale and proportion of cross-border e-commerce transactions in China showed a trend of increasing year by year. In 2018, the transaction scale of China's cross-border e-commerce reached 9.0 trillion yuan, accounting for 28.3% of the total import and export trade.

The penetration rate of China's cross-border e-commerce industry has increased year by year. In 2019, the penetration rate of China's cross-border e-commerce industry was 33.29%, an increase of 3.79% compared with 29.5% in 2018. The penetration rate of the cross-border e-commerce industry has increased year by year from 2014 to 2019, indicating that more and more consumers are willing to purchase products or services on e-commerce platforms.

3 The impact of the COVID-19 outbreak on China's cross-border e-commerce industry

3.1 Order volume is affected by the severity of the epidemic

As shown in Table 1, from 2019 to 2020, the average number of parcels sent by Chinese logistics companies to the world is as high as 60%. Due to the impact of the new crown epidemic, many parcels could not be sent out, especially in February 2020, which dropped significantly. The issue of e-marketing communication is currently very topical since today, in the time of the rapid development of digitization processes caused by the coronavirus crisis, it is an important part of business operation. As a result, it is important for businesses to consider which strategy to choose in order to achieve a long-term prosperity. Nowadays, the cooperation with influences is highly effective, since by means of sharing and recommendations on social platforms, they may present a given product or service to the public [14].

Table 1. Statistical table of the number of parcels sent by Chinese logistics companies from 2019 to 2020.

Time	Number of packages sent	Global Share
10.2019	29 600 000	66.56%
11.2019	34 750 000	64.55%
12.2019	35 150 000	62.59%
01.2020	33 680 000	60.36%
02.2020	30 360 000	50.54%
03.2020	34 650 000	63.23%
04.2020	40 970 000	56.72%
05.2020	34 330 000	48.79%
06.2020	24 850 000	45.37%
07.2020	18 980 000	42.32%

Source: Global Logistics Query Platform, [15].

China's cross-border e-commerce exporting countries are mainly distributed in the United States, Europe and other places, and the penetration and competitiveness of emerging markets are relatively low. Since February 2020 is the peak period of the COVID epidemic in China, European countries are worried that it will be affected, so the volume of express parcels has dropped significantly. The rebound began from March to April 2020, mainly due to the proper control of the epidemic in China and the absence of the spread of the epidemic abroad. By May, the outbreak of foreign epidemics began, and countries with severe epidemics began to issue policies to block cities, further hindering logistics, and the volume of express orders began to decline again.

3.2 Supply chain efficiency declines, hoarding of goods

The data on the sharp drop in the number of Chinese export containers in 2020 reflects the impact on the global supply chain due to the shutdown of production and insufficient production capacity in various parts of China, forcing the industry to face the danger of chain interruption of intermediate products.

As shown in Table 2, from January to June 2020, the number of China's container exports was 750,000, a year-on-year decrease of 43.7%. From January to February 2020, the shutdown of enterprises due to the epidemic reduced China's external supply, resulting in a significant drop in the number of export containers. Although there has been a rebound from March to June, the overall efficiency is still low.

Table 2. Data table of China's export container quantity from June 2019 to June 2020.

Time	Cumulative number of export containers	Cumulative year-on-year growth
06.2019	1 330 000	-22.5%
07.2019	1 600 000	-21.6%
08.2019	1 770 000	-24.7%
09.2019	1 930 000	-27.5%
10.2019	2 070 000	-29%
11.2019	2 250 000	-29%
12.2019	2 420 000	-29%
01-02.2020	210 000	-48%
03.2020	360 000	-44%
04.2020	490 000	-41.9%
05.2020	620 000	-44.7%
06.2020	750 000	-43.7%

Source: General Administration of Customs of China, [16].

3.3 International logistics blocked

The global outbreak of COVID-19 has caused logistics costs to continue to rise. The price of international long-term routes in Europe and the United States has risen from 20 yuan to 100 yuan, and the short-term Asian prices have risen from 10 yuan to 40 yuan. Global flights continue to shrink, and the current global air capacity has been reduced to less than 40%, the European air capacity has been reduced by 80%, and the US air capacity has been reduced by 70%.

During the epidemic in 2020, due to the reduction in the international aviation market capacity compared with normal times, the logistics capacity in the belly of passenger aircraft has dropped significantly, resulting in an increase in air freight time and a continuous increase in prices. Nearly 80% of cross-border e-commerce sellers said that international logistics prices are rising every day. Under the influence of factors such as declining air capacity, insufficient logistics manpower and regional border control, cross-border logistics has been seriously hindered.

4 Case study: The development of Ali International Station under the COVID epidemic

4. 1 Announcement of business support policy plans during the epidemic

On February 19, 2020, Ali International Station released a business support policy plan during the epidemic, which focuses on solving logistics, capital chain and cloud office problems.

a. In terms of logistics, Alibaba International Station Logistics cooperates with Cainiao and the ecology to provide customers with high-certainty full-link visualization logistics solutions.

b. To solve the capital chain problem, Alibaba International Station launched a corporate financial service support plan during the epidemic, and corporate finance and MYbank provided many financial rights and interests for merchants on the international station.

c. Regarding working from home, Alibaba International Station has cooperated with DingTalk to provide sellers with an exclusive online office support channel. In addition to the ordinary rights and interests of DingTalk, there are also a series of exclusive rights and interests.

4.2 New trade festival promotes digital new foreign trade

On March 2, 2020, the "2020 New Trade Festival" of Ali International Station was officially launched.

a. In terms of sources of business opportunities, Alibaba International Station is committed to bringing more and more accurate traffic to platform merchants. On the one hand, traffic channels are fully opened to create new growth points for online traffic, and on the other hand, accurate traffic is introduced to match industrialization and orientation for efficient conversion to ensure efficient conversion from traffic to orders. As shown in Figure 1, during the New Trade Festival in March, the order volume increased by 114% year-on-year.

b. In terms of contract performance assurance, Alibaba International has launched a series of measures in terms of logistics, supply chain and cash flow to ensure the certainty of contract performance for merchants.

c. In terms of operational efficiency improvement, Ali International Station has launched a series of measures aimed at improving the digital transformation and upgrading of small and medium-sized foreign trade enterprises.

As shown in Figure 1, Alibaba International Station achieved remarkable results in the New Trade Festival in March 2020. The total transaction volume increased by 167% year-on-year, the number of orders increased by 114% year-on-year, and the number of paid buyers increased by 77% year-on-year. China's new foreign trade format bucked the trend.

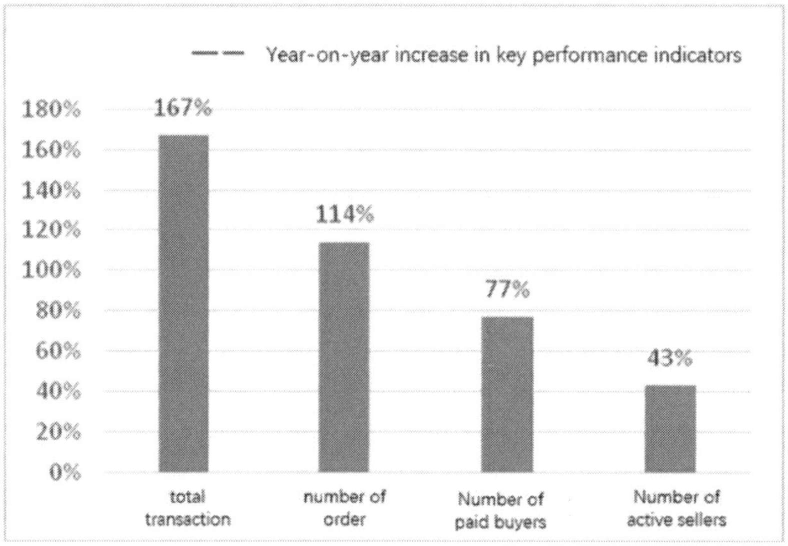

Fig. 1. Synchronous increase in key performance indicators of Ali New Trade Festival in March 2020 (unit: %).
Source: Alibaba.com, [17].

4.3 Open online fair to help transformation

a. Use live broadcast marketing to improve the quality of live broadcast content and accelerate the doubling of inquiries brought by merchants through live broadcast. Set up a short video special session to intuitively and three-dimensionally present the merchant's capabilities and product advantages.

b. With the help of the technical means of big data analysis, the precise matching of business opportunities between the supply and demand sides can be achieved.

c. Ali International Station launched a free audio conference function, which supports real-time one-to-one communication and negotiation between buyers and sellers, builds trust between buyers and sellers, and improves the efficiency of online communication.

The effect of merchants at the exhibition has improved significantly, and the number of active buyers per day of live broadcast merchants has increased significantly. According to the official data of Ali International Station, the average number of daily customer inquiries of the exhibitors at this online fair increased by 132.2% year-on-year, the average daily customer inquiries of live broadcast merchants increased by 129.9% year-on-year, and the number of paid orders increased by 243% year-on-year.

5 Path choices for China's cross-border e-commerce to deal with the covid-19 outbreak

5.1 Online exhibitions drive sales

At the moment of the epidemic, the exhibition industry is deeply affected. According to the "Research Report on the Convention and Exhibition Industry's Response to the COVID Epidemic" released by the China Convention and Exhibition Economic Research Association, nearly 3,500 domestic exhibitions and outbound exhibitions were affected by the epidemic in China from February to April 2020. The impact is more serious, and the impact of the conference industry is no less than that of the exhibition industry. Orders are the lifeline for enterprises to survive. If there is no exhibition, there will be no orders. Online exhibition will be a new trend for foreign trade enterprises to expand their business in the future.

The State Council's April 7 executive meeting decided that the 127th Canton Fair will be held online in mid-to-late June. Foreign trade enterprises should vigorously develop digital means to conduct online negotiations and use online exhibitions to attract customers. "Enterprises and exhibits are all online", without the limitation of traditional physical space, the number of products that enterprises can sell may increase significantly. In order to increase the exposure of products and enterprises, the exhibition recommends that enterprises make full use of the Google search engine commonly used in e-commerce, Facebook and other online community marketing and live broadcast delivery. An increasing number of retailers and service providers are aware of the importance of loyalty programmes as a tool to support sales [18].

As of June 24, 2020, the total export of the online Canton Fair was about 1,412.6 billion US dollars. It is worth noting that sellers who settled in Ali International Station created a year-on-year increase of 243% in the order volume at the online fair.

5.2 Accelerate the digital new foreign trade path

First of all, the cross-border e-commerce digital trade platform has entered the stage of digitalization of the industrial chain from the stage of digitalization of marketing and digitalization of transactions, that is, from the single online display function of the year, to gradually satisfying the basic trade demands of order placement, payment, and logistics, and then it has developed into a one-stop comprehensive foreign trade service platform that can support customs clearance, overseas warehousing, foreign exchange settlement and tax refund, and agency operations.

Secondly, the difficulty of forming a global enterprise will be reduced rapidly, and "zero-based globalization" will be finally realized. With the help of the digital platform of cross-border e-commerce, even companies lacking foreign trade experience do not need to spend a lot of money and time to find various service providers, quickly start foreign trade business, and establish a global marketing network, supply chain system and contract performance channel. Most successful companies have established effective marketing communication as a useful tool for selling their products to customers [19]. Not only the improved visibility into marketing effectiveness has helped companies to better prioritize their efforts, but it also has improved the quality and precision of the marketing resources demanded by the sales force. Finding the appropriate key metrics should contribute to the evaluation of marketing effectiveness [20].

5.3 Further improvement of international logistics channels

The COVID-19 epidemic has made cross-border e-commerce companies realize the outstanding importance of logistics. Insufficient international transportation capacity, limited overseas warehouses, and poor logistics channels are important factors that inhibit the development of cross-border e-commerce in China in the short term.

First of all, Chinese cross-border e-commerce companies actively optimize international logistics routes to ensure product delivery, and open overseas warehouse services to the vast number of cross-border e-commerce sellers to ease the pressure of Amazon FBA to restrict the warehousing of goods.

Secondly, in terms of air transport, we actively carry out the "passenger-to-cargo" business, temporarily transforming passenger planes into cargo planes, and continuing to send epidemic prevention materials and cross-border e-commerce products.

Finally, China's cross-border logistics supply chain enterprises should fully release their vitality and start to build independent international logistics channels in an all-round way, which will help reduce China's cross-border e-commerce's dependence on foreign logistics enterprises and improve their ability to resist risks.

6 Conclusion

The aim of this paper was to recognize and subsequently determine possible problems that may be an obstacle to the future development of e-commerce of companies. Based on the results achieved, the conclusions presented below can be stated.

Under the government's promulgation of many cross-border e-commerce industry support policies, cross-border e-commerce enterprises hold online exhibitions and other measures to help traditional small and medium-sized enterprises and foreign trade merchants to further transform to online. By using the new digital foreign trade model, it can precisely match the information and needs of buyers and sellers, thereby saving operating costs and improving efficiency. During the epidemic, the problems existing in the cross-border e-commerce industry have been highlighted, and many improvements have been made in the global supply chain and logistics system to help sellers solve the problems of contract performance, stockpiling and cash flow. Although the COVID-19 epidemic has had an inevitable impact on the cross-border e-commerce industry, with the joint efforts of the government, cross-border e-commerce platforms and businesses, the difficulties were overcome and the crisis was finally turned into an opportunity. This opportunity is precisely the development of e-commerce to global dimensions, which has helped not only to simplify but also to speed up cross-border trade. This also had a positive impact on the incentive to modernize and improve global logistics. This has a positive effect on the development and further functioning of road, air and ship transport.

Further research in this area could focus on evaluating the impact of the Covid-19 pandemic on the development of e-commerce in other countries. The topic of e-commerce itself brings many research opportunities, as this industry is relatively young and is currently starting to come to the fore. Further research and the shift of industry and its economy towards Industry 4.0 are also related to this.

References

1. K. Hayakawa, H. Mukunoki, S. Urata, Can e-commerce mitigate the negative impact of COVID-19 on international trade? *JER* (2021)
2. A. Abudureheman, A. Nilupaer, Optimization model design of cross-border e-commerce transportation path under the background of prevention and control of COVID-19 pneumonia. *Soft Comput,* **25,** 12007-12015 (2021)
3. R. Nacar, K. Özdemir, E-ticaret pazaryeri web Sitelerinin Dinamikleri: Covid-19 pandemi sürecindeki değişim. *Eskişehir Osmangazi Üniversitesi İktisadi ve İdari Bilimler Dergisi,* **16**(1), 205-226 (2021)
4. T. Chen, Y. Qiu, B. Wang, J. Yang, Analysis of effects on the dual circulation promotion policy for cross-border e-commerce B2B export trade based on system dynamics during COVID-19. *Systems,* **10**(1), 13 (2022)
5. A. A. Agus, G. Yudoko, N. Mulyono, T. Imaniya, E-commerce performance, digital marketing capability and supply chain capability within e-commerce platform: Longitudinal Study before and after covid-19. *International Journal of Technology,* **12**(2), 360 (2021)

6. T. Kawasaki, H. Wakashima, R. Shibasaki, The use of e-commerce and the COVID-19 outbreak: A panel data analysis in Japan. *Transport Policy,* **115**, 88-100 (2022)
7. H. Guo, Y. Liu, X. Shi, K. Z. Chen, "The role of e-commerce in the urban food system under COVID-19: lessons from China". *China Agricultural Economic Review*, **13**(2), 436-455 (2021)
8. J. Horak, Sanctions as a catalyst for Russia's and China's balance of trade: Business opportunity. *Journal of Risk and Financial Management*, **14**(1), 36 (2021)
9. A. U. Din, H. Han, A. Ariza-Montes, A. Vega-Muñoz, A. Raposo, S. Mohapatra, The impact of covid-19 on the food supply chain and the role of e-commerce for food purchasing. *Sustainability*, **14**(5), 3074 (2022)
10. L. Zhiyong, The impact of the novel coronavirus pneumonia epidemic on cross-border e-commerce and countermeasures. *Computer and Network*, **4**, 9-10 (2020)
11. D. Gajdosikova, K. Valaskova, T. Kliestik, V. Machova, COVID-19 Pandemic and its impact on challenges in the construction sector: A case study of Slovak enterprises. *Mathematics*, **10**(17), 3130 (2022)
12. Q. Ding, H. Zhao, Study on e-commerce logistics cost control methods in the context of COVID-19 prevention and control. *Soft Comput*, **25**, 11955-11963 (2021)
13. P. Domanizova, F. Milichovsky, K. Kuba, Business models,strategy and innovation in the new world of digitization. *Littera Scripta,* **13**(1), 1-15 (2020)
14. B. Lukach, Y. Kostiuk, Current trends in e-marketing – empirical analysis of selected social platform. *Littera Scripta,* **15**(1), 56-67 (2022)
15. Global Logistics Platform (online) Available at: https://old.logisview.com/main.do?locale=en (10.11.2022)
16. General Administration of Customs of China (online) Available at: http://english.customs.gov.cn/ (10.11.2022)
17. Ali New Trade Festival in March 2020 (online) Available at: www.alibaba.com (11.11.2022)
18. M. Zilinska, J. Vrbka, E. Kalinova, Does the implementation of loyalty programmes lead to gaining a loyal customer? *Ad Alta-Journal of Interdisciplinary Research*, **11**(1), 392-396 (2021)
19. K. Mikulova, M. Vitek, Marketing Communication of a business subject on the consumer electronics market during covid-19 pandemic. *Littera Scripta,* **15**(1), 68-81 (2022)
20. F. Milichovsky, I. Simberova, Marketing effectiveness: Metrics for effective strategic marketing. *Inzinerine Ekonomika-Engineering Economics,* **26**(2), 211-219 (2015)

Return on sales in the Czech building industry? Does firm size matter?

Tomáš Krulický[1,*] and *Dominik* Potužník[1]

[1]Institute of Technology and Business in České Budějovice, School of Expertness and Valuation, Okruzni 517/10, 37001 Ceske Budejovice, Czech Republic

Abstract. The presented paper explores the link between firm size and returns on sales (RoS) in Czech building companies. We examined how different business sizes relate to the return on sales, considering staff numbers and the RoS movement over a period. The CRIBIS database of CRIF Company - Czech Credit Bureau provided relevant data from 2016 to 2020, including profitable and active building enterprises. The industry analysis involves an empirical approach using industry structure analysis, return on sales analysis and observing its movement from 2016 to 2020. The industry predominantly encompasses small firms of fewer than 25 employees, exhibiting dramatic variations of the average return on sales over the period. Large companies exceeding 250 employees show complete conformity, yet their numbers are small in the construction sector. The correlation analysis revealed that the business size, i.e., staff numbers, is independent of the return on sales. The submitted work suggests a new size classification according to total assets and sales.
Keywords: Cost of capital, size premium, discount rate, valuation

1 Introduction

What are the sizes of building enterprises in the Czech Republic? Must the companies have many employees to be profitable? Experts have been in dire straits answering these questions, given the sluggish demand in the construction industry. Building companies in the Czech Republic are plentiful and lack a qualified workforce, which means fierce competition in the market. The enterprises want to know if size matters in their profitability. The current global economy, according to some authors defined as a new economy, is characterized as turbulent. Emphasis is placed on generating value added [1].

The building industry is imperative for the national economy and socio-economic development, including housing development, facilities and infrastructure. [2] argue that the construction sector involves capital flows, individuality, workforce and requirements. Businesses are the driving force of economies in today's globalized world, especially in developing countries. These companies associate buyers with sellers, but do not participate directly in the ownership or storage of goods, they make revenues through commissions.

*Corresponding author: krulicky@mail.vstecb.cz

Globally, the goal of any business is to increase the value of an enterprise. To determine the value of the business, traditional financial analysis proved to be inadequate. The need for application and use of new indicators that more accurately measure the efficiency of both the processes and the elements that generate the value caused great interest in the so-called value generators. Value generators refer to activities and capabilities that increase profitability, reduce risk, and promote growth in the company [3].

[4] claim the building industry contributes to the national economy by planning, developing, modifying, and repairing buildings or facilities. The building industry allows investments in social aspects (sport, healthcare, housing etc.), energetics, transport, agriculture and production. The arrival of the pandemic had a negative impact on a number of aspects of business in the construction industry and affected the financial and economic situation of companies in the construction industry in the Slovak Republic. Since this sector is characterized by slower reactions to changes in the economy, the most significant impacts will be even more noticeable in the future, which is why it is also considered a key sector of the Slovak economy [5].

The firm size depends on many external and internal factors. Most enterprises develop their activities relying on sales, profits, staff numbers or the size of premises. [4] assume that many companies competing in ever-changing sectors must expand (increased productive capacity, convenient location, market share etc.) to survive.

Experts have extensively considered the pros and cons of firm size [6] arriving at various conflicting conclusions.

Through time series analysis, it is possible to obtain significant statistics and other necessary data characteristics. Prediction of time series allows predicting future values based on previously observed values. The exact prognosis of the time series is very important for a number of different areas, such as transport, energy, finance, economics, etc. [2].

2 Research aims

The paper explores a link between staff numbers (firm size) and return on sales of Czech building companies from 2016 to 2020.

The study aims to find if the firm size in the Czech building sector reflects its return on sales.

We formulated two research questions:

1. What is the share of different firm size categories (staff numbers) in the Czech building sector between 2016 and 2020?
2. How much is firm size impactful on the return on sales in the Czech building sector between 2016 and 2020?

3 Literature research

There have been many theories why some firms are more profitable than others, analyzing decisive business performance variables. Yet scientists and professionals still cannot hit the nail on the head.

Experts on industrial economy, business organization and finance consider a firm size imperative for interpreting the company's profitability [7].

Google searches, interviews with investors and financial analysts or watching news suggest a deep-seated belief that profits scale with the business size [8].

Staff numbers reflect the firm size better than sales or assets, rendering the question of profitability disputable [9]. [10] say that the profitability analysis is an integral part of the financial analysis of a company. Profit is one of the most commonly used indicators of a company ' s performance and is used to evaluate a company ' s historical performance. Given their importance, it examines how the available information on a company ' s profit can be used to increase the profitability.

[11] conducted statistical tests to draw reasonable conclusions, including return on sales, capital structures and asset liquidity. The survey involved 61 large enterprises in North America, Europe and Asia, revealing no significant correlation between the firm size and profit rates.

[2] focused on size determinants (total assets, sales and staff numbers), stimulating the performance of 51 small and medium-sized enterprises in Portugal between 1999 and 2003. The dynamic panel regression model showed that all firm size measurements significantly correlate with business performance, including the operating income divided by total assets for SMEs. The authors compared the findings of SMEs with large companies, applying the regression to large firms. The results indicated no link between the firm size and the return on sales.

Empirical research in various sectors, regions and periods deals with profit ratios, revealing that profit-making companies reinvigorate a healthy economy. [7] reached mixed results when analyzing firm size. Although cross-sectional regression proved no significant correlation between the return on sales and total sales and assets, the time series analysis showed a close interrelationship. [12] used KGI and KPI indicators to manage performance, further applied the Balanced Scorecard to micro and small and medium-sized enterprises.

The graphical analysis of 3,139 Czech building companies between 2011 and 2015 indicated a higher return on sales in large enterprises [6].

Gibrat's law contrasts with empirical studies, stating that the firm's growth rate is independent of its size. [13] support this theory, analysing the growth trend of small, medium-sized and large enterprises with high profits. The data reflected 124 building companies registered in Malaysian BURSA between 2003 and 2010. The research revealed that the growth contributed to the return on sales in small and medium-sized construction companies but indicated no involvement in large enterprises. A crisis may arise in any business, so it is essential to focus on any problems in order not to endanger the company viability or to be out of business activity [1].

4 Data and methods

The CRIBIS database (CRIF Company) - Czech Credit Bureau provided the necessary data for the period, including balance sheets, income statements, return on sales, assets, staff numbers and business location. Our analysis observes only the data from active and profit-making building companies, ignoring inactive, loss-making or winding-up firms. The data span 2016 - 2020, falling into the 'F' Section of the CZ NACE Economic activities. The section involves specialized and non-specialized construction activities.

Tables in Excel provide a list of data for each new year, including a table for firm size categories. Table 1 suggests the given business sizes.

Table 1. Firm size categories.

Firm size	Staff numbers
Small	1-24
Medium-sized	25-249
Large	250 and more

Source: Own processing.

Then, we explored the firm's involvement in the categories from 2016 to 2020. Table 3 suggests the proportions, including an overview of the firm size trend over the period. Graph 1 below illustrates the summary.

The Excel lists propose the analysis of return on sales for each year, comprising a tabular overview of the average return on sales ('AVERAGE'), minimum ('MIN') and maximum ('MAX') profitability. The time series analysis generates Graph 2, showing the profitability trend throughout firm sizes from 2016 to 2020.

Correlation analysis explores if the firm size contributes to the return on sales, using tabular correlation coefficients for each year in Excel lists. The table of coefficients contains a return on sale value for each analyzed enterprise in one column and the firm size in the second, marked with a rate from Table 2 below.

Table 2. Values of the firm size categories.

Firm size	Value
Small	1
Medium-sized	2
Large	3

Source: Author's elaboration.

Then, the CORELL function in Excel allows replacing matrix one with values of the firm size categories and matrix two with business profitability rates, achieving a correlation coefficient for the given year. The table suggests the correlation coefficients for 2016 - 2020.

5 Results

Observing the laid-down criteria, the data set provided the number of Czech building enterprises from 2016 to 2020, filtering the businesses by small, medium-sized and large. Table 3 shows the total of the companies over the period, including their involvement in the category.

Table 3. The firms' involvement in the categories in 2016-2020.

Year	Small firm	Medium-sized firm	Large firm	Total
2016	6,635	782	37	7,454
2017	7,022	989	52	8,063
2018	7,242	1,019	55	8,316
2019	6,598	1,010	58	7,666
2020	4,596	880	55	5,531

Source: Author's elaborations from the CRIBIS database.

Table 3 suggests a slight increase in the total of businesses in the first three years. Years 2019 and mainly 2020 saw a decline of about one-third, 28%. Most enterprises complied with the criteria in 2018, peaking at 8,316. The table further shows the closest involvement of small companies (up to 24 employees), attaching the lowest numbers to large firms (over 249 employees). Figure 1 shows that small companies saw a predominant downward trend in meeting the criteria.

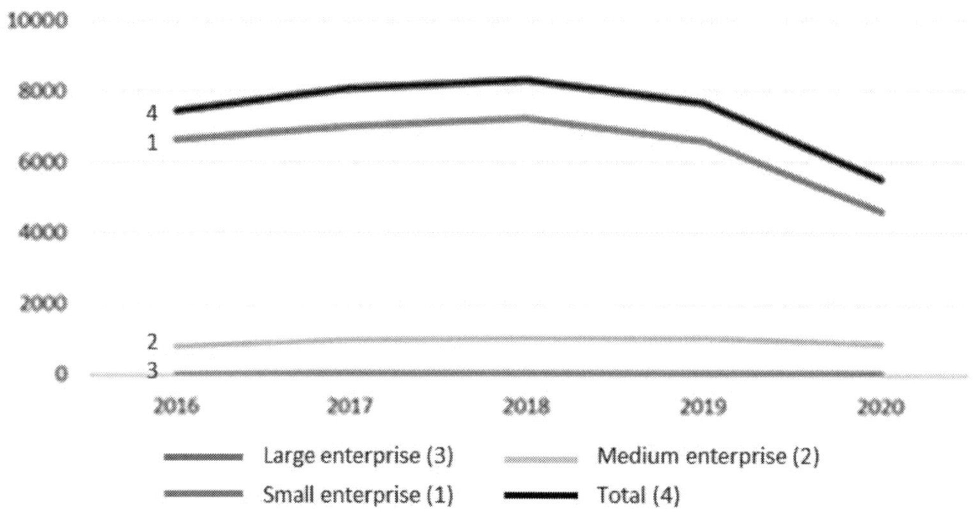

Fig. 1. The trend in enterprise numbers in 2016-2020. Source: Author's elaboration from the CRIBIS database.

We then analysed the return on sales for firm size categories over the period, calculating AVERAGE, MIN and MAX values in Excel. Tables 4, 5, 6, 7 and 8 propose the obtained figures for separate years, respectively.

Table 4. Analysed return on sales (RoS) for 2016.

Firm size	Small	Medium-sized	Large
Average RoS	1.7	0.06	0.1
Max RoS	8.846	1.17	1.23
Min RoS	-5.29	-0.02	0.01

Source: Author's elaboration from the CRIBIS database.

The year 2016 saw small enterprises top the average return on sales to 1.7 and medium-sized companies fall to 0.06. Large firms peaked at 0.1.

Table 5. Return on sales for 2017.

Firm size	Small	Medium-sized	Large
Average RoS	0.55	0.06	0.15
Max RoS	1.498	1.33	5.24
Min RoS	-0.18	-0.13	0.01

Source: Author's elaboration from the CRIBIS database.

The same scenario unfolds for 2017, involving slight modifications in figures. Small enterprises peaked at 0.55, leaving behind medium firms with 0.06, indicating the lowest values. Large corporations reached 0.15.

Table 6. Return on sales for 2018.

Firm size	Small	Medium-sized	Large
Average RoS	0.3	0.32	0.11
Max RoS	837.9	254.3	2.81
Min RoS	-1.19	-0.02	0

Source: Author's elaboration from the CRIBIS database.

2018 launched medium-sized enterprises to the first place with 0.32, narrowly defeating small firms with 0.3. Large companies reached 0.11, indicating the lowest return on sales, dropping by 0.4 from the previous year.

Table 7. Return on sales for 2019.

Firm size	Small	Medium-sized	Large
Average RoS	0.46	0.07	0.11
Max RoS	974	0.83	3.1
Min RoS	-1.1	-0.02	-0.04

Source: Author's elaboration from the CRIBIS database.

In 2019, small companies recovered the first position, peaking at 0.46, indicating an improvement of 0.16. Medium-sized firms suffered a relapse to the last place, reaching 0.07 of the average RoS, falling by 0.25. Large organizations were on the level as the previous year, achieving 0.11.

Table 8. Return on sales for 2020.

Firm size	Small	Medium-sized	Large
Average RoS	0.25	0.07	0.12
Max RoS	113	1.49	3
Min RoS	-0.26	-4.67	0.01

Source: Author's elaboration from the CRIBIS database.

Although small firms received the highest average returns in 2020, the profit slumped by 0.21. The scenario was par for the course with medium-sized enterprises, indicating 0.07. Large companies moved up to 0.12, making an improvement of 0.1. Graph 2 suggests the trend of the average RoS from 2016 to 2020 for all size categories.

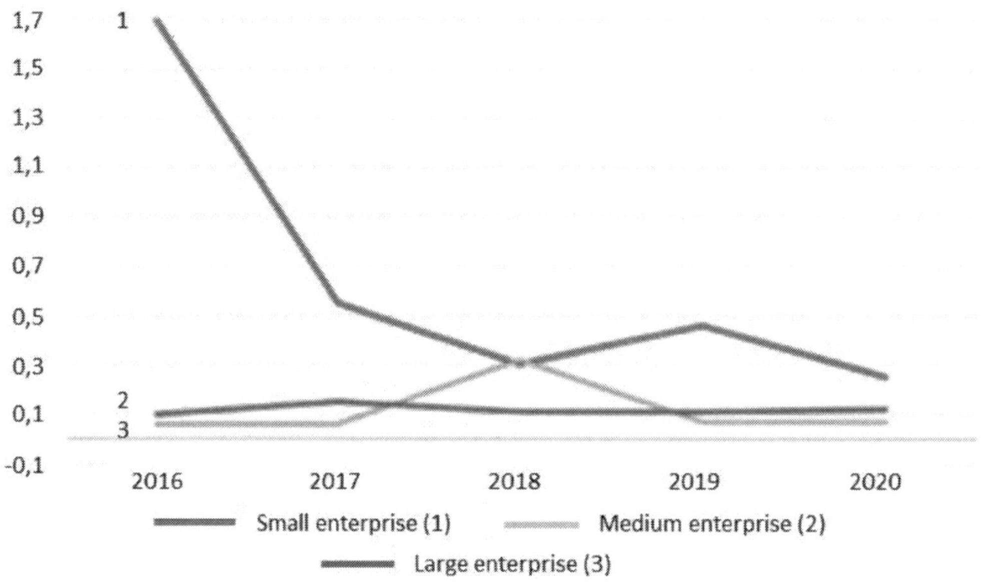

Fig. 2. Average RoS in 2016-2020. Source: Author's elaboration from the CRIBIS database.

Figure 2 shows large enterprises as the most stable entities regarding the average RoS over the period. Medium-sized firms were constant as well, significantly deviating only in 2018. On the flip side, small organizations marked dramatic departures throughout the period.

Table 9. Correlation coefficient for 2016-2020.

Year	Correlation coefficient
2016	0
2017	-0.01
2018	0
2019	-0.01
2020	-0.02

Source: Author's elaboration from the CRIBIS database.

The monitored period did not see the correlation coefficient getting over zero. Years 2017, 2018 and 2020 witnessed the coefficient drop slightly below zero, i.e. max -0.02 in 2020. The coefficients suggest that return on sales does not depend on the firm size of the building company.

6 Discussion

We formulated two research questions:

1. What is the share of different firm size categories (staff numbers) in the Czech building sector between 2016 and 2020?

We found that the number of analyzed enterprises that met our research criteria increased until 2018, topping 8,316 firms in the same year. The aggregate notably declined between 2019 and 2020, where we can blame the economic restrictions and crisis triggered by the Covid-19 pandemic. Many companies wound up or became loss-making, which was inconsistent with our parameters. In times of economic crisis, leasing companies can be an alternative way to obtain a loan. However, their value is constantly changing during economic cyclical events [14].

The analysis showed that the Czech building industry mostly involved small enterprises (less than 25 employees) between 2016 and 2020. Small-scale firms comprised more than 80% of all corporations in the sectors over the period. On the flip side, large firms (more than 249 employees) did not exceed 1% in total within the period.

2. How much is firm size impactful on the return on sales in the Czech building sector between 2016 and 2020?

According to our analysis, small enterprises show the sharpest RoS fluctuations, indicating the steadiest return levels in large companies over the period. Medium-sized firms demonstrated profitability divergence only in 2018, keeping constant values for the rest of the period.

Correlation analysis suggests that the RoS of the examined enterprises does not relate to the firm size. At first, the correlation coefficient was 0, rising slightly above zero in the last year.

Our findings are close to those of [4] and [2] who did not prove any significant correlation between the firm size and return on sales. The authors claim that only the total assets are impactful on corporate profitability.

7 Contribution

Experts still cannot see eye to eye on the matter. What is the involvement of firm size in the Czech building industry? Do the enterprises have to have high staff numbers to be profitable? According to this study, the building industry between 2016 and 2020 mostly comprised small firms (low staff numbers). Large companies (over 249 employees) did not exceed one per cent of the analyzed enterprises.

According to the profitability and correlation analysis, staff numbers do not affect the return on sales, indicating that even small firms can generate profit. Small companies are also receptive to violent economic fluctuations, whereas large organizations maintain remarkable consistency. We may argue that small firms are highly susceptible to changes, while large corporations go native dramatic scenarios like the Covid-19 pandemic or increased competition.

The study contributes to existing building companies, informing on the competition in their size category. The work also helps new Czech construction enterprises explore the total of firms in the sector and their involvement according to the staff numbers, advising on the average return on sales in the size categories.

8 Conclusion

We analyzed Czech building companies over 2016 - 2020, focusing on staff numbers and return on sales. The research explored if the firm size impacts the profit in the building industry over the period.

We also defined two research questions: What is the share of different firm size categories (staff numbers) in the Czech building sector between 2016 and 2020, and how much is firm size impactful on the return on sales in the Czech building sector between 2016 and 2020?

The chapter 'Results' involved size categories between 2016 and 2020, demonstrated in a graph. Small enterprises, with fewer than 25 employees, had the strongest representation. We calculated the average return on sales for all years, assigning each size category the minimum and maximum profit rate. Small companies witnessed the most dramatic fluctuations, whereas large corporations stayed remarkably constant. The last step included a possible link between the firm size and return on sales, revealing n connection between staff numbers and profitability.

The study contributes to existing building companies, advising them on the competition in their size category. For new Czech building startups, the work may be instrumental in disclosing the total of enterprises in the sector and their size categories (staff numbers). The study is deficient in having a lack of data on firms that could not be involved in the research.

Further analysis may focus on categorizing companies according to size, considering total assets or sales and exploring the results.

References

1. P. Suler, V. Machova, Financial Stability in Central European Countries: Case of The Visegrad Four. *7th International Conference on Education and Social Sciences (INTCESS 2020),* 1218-1223 (2020)
2. Z. S. Serrasqueiro, P. M. Nunes, Performance and Size: Empirical Evidence from Portuguese SMEs. *Small Business Economics,* **31**(2), 195 – 217 (2008)
3. D. Kim, S. C. Lin, T. C. Chen, Financial structure, firm size and industry growth. *International Review of Economics,* **41**, 23-39 (2016)
4. J. Niresh, T. Velnampy, Firmsize and profitability: A study of listed manufacturing firms in Sri Lanka. *International Journal of Business and Management*, **9**(4), 57-64 (2014)
5. J. Horak, V. Machova, Business Enterprises and Generators of their Value. *18th International Scientific Conference on Globalization and Its Socio-Economic Consequences,* 547-554 (2018)
6. B. Groda, Is the size of enterprise in the construction industry an important prerequisite for its success? *Innovative Economic Symposium (IES) on Strategic Partnership in International Trade,* **39** (2017)
7. A. Kumar, V. Srivastava, M. Tabash, D. Chawda, Profitability determinants of infrastructure public private partnerships (PPPs): empirical evidence from Indian data. *Journal of Financial Management of Property and Construction*, **27**(1), 91-111 (2022)
8. V. Machova, J. Marecek, Estimation of the development of Czech Koruna to Chinese Yuan exchange rate using artificial neural networks. *Innovative Economic Symposium 2018 - Milestones and Trends of World Economy (IES2018),* **61** (2019)
9. M. Dogan, Does firm size affect the firm profitability? Evidence from Turkey. *Research Journal of Finance and Accounting*, **4**(4), 53- 59 (2013)
10. J. Marecek, V. Machova, Importance of Analyzing the Development of the National Economy in the Context of a Global World. *18th International Scientific Conference on Globalization and Its Socio-Economic Consequences,* 2629-2636 (2018)
11. C. Y. Yee, C. Y. J. Cheah, Interactions between Business and Financial Strategies of Large Engineering and Construction Firms. *J. Manag. Eng*, **22**, 148–155 (2006)
12. M. Psarska, S. Haskova, V. Machova, Performance Management in Small and Medium-Sized Manufacturing Enterprises Operating in Automotive in the Context of Future Changes and Challenges in SR. *Ad Alta-Journal of Interdisciplinary Research,* **9**(2), 281-287 (2019)
13. D. Rasiah, D. Tong, P. K. Kim, Profitability and Firm Size-Growth Relationship in Construction Companies in Malaysia from 2003 to 2010. *Review of Pacific Basin Financial Markets and Policies,* **17**(3), 1-19 (2014)
14. J. Kucera, P. Suler, Development of Economic Value Added (Eva) of Leasing Companies Active in The Czech Republic in the Years 2005-2019. *Ad Alta-Journal of Interdisciplinary Research,* **10**(2), 218-223 (2020)

The SERVQUAL instrument to measure service quality in higher education – A case study

Ana Rolo[1,*], *Rui* Alves[2], *Margarida* Saraiva[3], and *Gracieth* Leandro[4]

[1]CICE, Instituto Politécnico de Setúbal, Portugal
[2]Instituto Politécnico de Setúbal, Portugal
[3]Universidade de Évora, Portugal
[4]Universidade Agostinho Neto, Luanda

Abstract. Students and employers, are the "clients" of Higher Education Institutions (HEI), so it is important to obtain their feedback on the quality of HEI and the courses they attend. This article aims to listen to students' opinion as "clients" of higher education, based on their expectations and perceptions, to measure the quality of the service of HEI. The results from empirical research were presented on the measurement of expectations and perceptions to assess the quality of services provided by a HEI perceived by students, based on observations obtained through survey. Quantitative methodology and data analysis techniques were applied. The SERVQUAL questionnaire was filled by 271 students who attended the 1st, 2nd and 3rd year of the five undergraduate courses taught by the institution, in the academic year 2021/2022. The values obtained for the expectations were very high which translated into negative deviations (Gaps). Thus, we can conclude that ESCE students are very demanding, have high expectations or are very demanding when they classify perceptions by assigning them low values.
Keywords: Higher Education, Service Quality, SERVQUAL

1 Introduction

Schools have generally been outside the concepts and theories of management. However, a school is an organization, with a similar structure to the one of the public companies.

In the current political and economic context, characterizes by cuts in the budget and shortage of students, Higher Education Institutions (HEIs) directors started to give greater attention to the problem of quality and to act as a company in the global market.

Thus, HEI, which in the past did not have to concern to attract students (the number of students applying for Higher Education was always much higher than the vacancies available, even considering competition from institutions private) have had to readapt to a very different reality in recent years. The decrease in the birth rate that has been observed in Portugal (Figure 1), has contributed to the decrease in the student population applying for higher education, so, there is more supply than demand according to the market laws. In another hand, HEI are integrated in a competitive market, and they face challenges due to the evaluation of the results of the services of education and training that they offer. The school

* Corresponding author: ana.rolo@esce.ips.pt

increasingly needs to be competitive to make the difference from its competitors through the quality of education and reputation of the institution. Prestige is constructed on quality objectives. There is a need for urgent changes in the education system and schools. This can be obtained by the implementation of a new model of management centered on the quality of services supplied to customers.

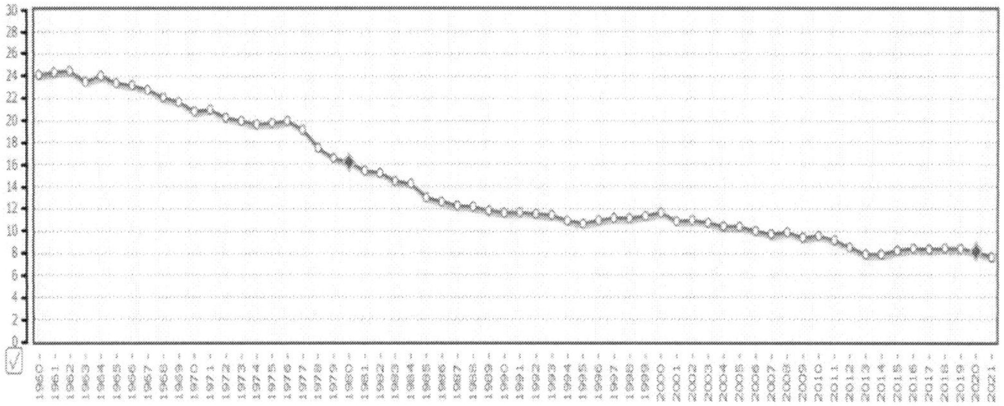

Fig. 1. Birth Rate Evolution (1960-2021).

Source: https://www.pordata.pt/Portugal/Taxa+bruta+de+natalidade-527.

The Portuguese higher education system is now characterized by: high number of public and private institutions; excessive numbers of graduation courses; absence of a relationship between educational activities and academic research and an inappropriate name of courses for marketing purposes.

In this context, quality plays a key role in differentiating the offer and will influence students' choice, forcing HEIs to prioritize these aspects, as they cannot remain outside the concepts and theories of marketing and management.

Another challenge relates to the evaluation of the results of the education and training services they offer, by external entities. To this end, structures were created with the aim of evaluating and monitoring the activities carried out by HEIs, such as the Higher Education Assessment and Accreditation Agency (A3ES). The main objective of A3ES "is to improve the quality of performance of higher education institutions and their study cycles and ensure compliance with basic requirements" (in http://www.a3es.pt).

This article aims to listen to students' opinion as "clients" of higher education, based on their expectations and perceptions, to measure the quality of the service of HEI and to identify characteristics of services that are important to students in order to provide guidelines for ESCE's strategic planning. By having access to this information, ESCE can accomplish its mission more efficiently.

Research Questions:

1. Does quality in education have 5 dimensions as is out forward Parasuraman et al. to the quality of services?
2. Do the dimensions have the same relative importance?
3. Are the dimensions identified as most important by the students included as priorities in the strategic planning of ESCE – IPS?
4. What are the most important determinants of overall service quality in ESCE?

The paper is organized as follows: Section 1 describes the theoretical basis and presents SERVQUAL Model as well as the description of the Expectations versus Perceptions as basis of this construct. Section 2 - Data and Methods, presents the methodological approach applied in the analysis of data gathering and the description of the case study. Section 3 shows some of the results and introduces a discussion of findings. At the end, Section 4, presents the identified limitations and future research directions.

2 Theoretical basis

In this section, we introduce a literature review os the SERVQUAL Model and a brief description about other models that can be applied to measure HEI service quality.

2.1 SERVQUAL model

SERVQUAL is a multidimensional research instrument designed to measure service quality by capturing respondent's expectations and perceptions along five dimensions of service quality [1, 2].

According to the same authors [2, 3], the main objective of the SERVQUAL model is to design measurement scales that allow evaluating the quality of services based on the difference between expectations and perceptions of customers. The service expected by the customer is the standard or "benchmark" against which experiences are compared; as performance exceeds expectations, service quality increases and vice versa.

The measurement of quality in services is characterized by being complex, abstract, multifactorial (influenced by multiple factors with different weights) and intangible and is defined as the degree of adjustment between the characteristics of the service and the attributes valued by the customer. These characteristics are not always easy to translate, due to the nature of the service itself, which enjoys the following characteristics: intangibility (it has no physical existence, cannot be viewed or touched, which prevents the establishment of precise specifications), perishability (production and consumption occur simultaneously), inseparability (the customer participates in the production process, being able not only to participate passively, but also as a co-producer of the service) and variability (their performance varies from person to person, from consumer to consumer and from day to day).

If we intend to define service, [4] defines it as a process constituted by a set of intangible activities that, generally but not always, are carried out through interaction between customers and resources of the entity providing the service (human, physical and computer resources), activities that are provided as solutions to customer problems.

[5], states that a service is a psychological and fundamentally personal result. For this author, the purpose of listening to customers about what they are experiencing is to find out what to improve to keep customers or get new ones. The customer's participation is an excellent opportunity to listen to him; [6] states that all major service quality studies agree that service quality is so subjective that it can only be measured in terms of what consumers want or define as quality.

Expectations versus perceptions

The SERVQUAL model measures the expectations - what the customer expects to get from a given experience or service and constitute a frame of reference or basis for the consumer to evaluate the performance of each service. Customer expectations, when correctly identified, are valuable information for organizations.

The model also measures customers' perceptions of the service they experience, which reflect their experience.

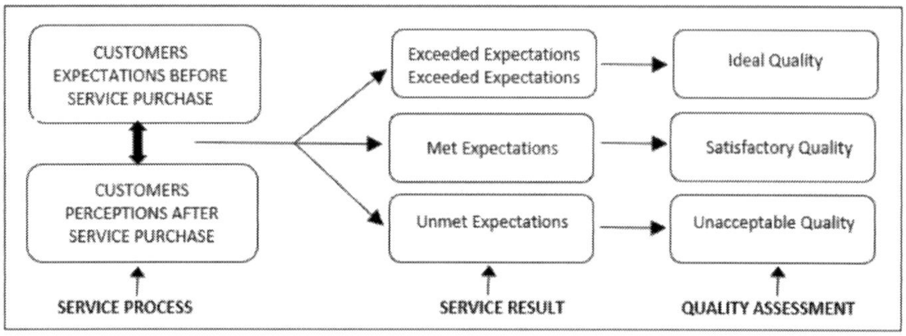

Fig. 2. SERVQUAL Model.

Source: Own processing.

As observed in Figure 2, Service Process is measured by the difference between customers' expectations before service experience and their perceptions after service experience. The result of this difference can be positive, null or negative. If it is positive, the service exceeded the customer's initial expectations and customers classify service quality as ideal; if it is null, quality is classified as satisfactory; if the result is negative, the perception of service quality has not the best and didn't achieve customers' expectations; in this case quality is considered unacceptable.

2.2 Other models

There are other models to measure the quality of services in education, such as HEdPERF (Higher Education Performance-only) as used by [7]. In their study, the authors proposed 41-item instrument based on the SERVPERF model. The purpose of this scale is to measure service quality specifically in the higher education sector, as according to the author, the generic scales presented previously may not be adequate for this purpose. The results were crucial because previous studies have produced scales that bear a resemblance to the generic measures of service quality, which may not be totally adequate to assess the perceived quality in higher education. Previous research focus only on academic's quality and gives few attention to non-academic aspects of the educational experience.

The European Foundation for Quality Management (EFQM) is another model applicable to measuring quality in services. This excellence model has been widely used by Higher Education Institutions (HEIs), but knowledge about its application in the context of HEIs is still limited [8, 9].

This research [10, 11] presents the comparison of the SERVPERF model with other generic instruments. determine which of these three instruments has superior measurement capacity in terms of one-dimensionality, reliability, validity and explained variance of service quality. The HEdPERF, designed exclusively for higher education, is empirically tested as the most comprehensive and industry-specific scale.

3 Data and methods

In order to test the SERVQUAL model and answer the listed research questions, the questionnaire was applied to ESCE (School of Business Administration) students at the Polytechnic Institute of Setúbal.

3.1 Case study – ESCE

ESCE was founded in 1994, and from the beginning sought an opening strategy that would allow ESCE to offer its students, courses that would provide training to integrate them into the business world. Although the school did not have it at the time, a culture of quality was being built in the continuous search for higher levels of performance and this has been its aim in recent years. Being competitive is having the ability to respond to market needs. ESCE knows that the differentiating factor between HEIs is related to quality improvement, which is why it has institutionalized a set of procedures capable of permanently evaluating the pedagogical functioning of the subjects and the quality of the teaching-learning process.

3.2 Methodology

The case study is one where the modified SERVQUAL Model was applied to ESCE to measure the students' expectations and perceptions of the service quality carried by this institution. A quantitative methodology was used, and data analysis techniques were applied to the available elements, with the aid of the SPSS statistical software. The SERVQUAL questionnaire (Appendix 1), using a likert scale of 7 points, was completed by a stratified sample of 271 students, attending the 1st, 2nd and 3rd years of the five undergraduate courses offered by ESCE in the 2021/2022 academic year. The questionnaires were distributed by email to students using the Survey Monkey platform. The response rate was 26%.

This SERVQUAL questionnaire includes a third section, that measures the relative importance of the five dimensions to the students. These scores are then used to weight the perceived service quality measure for each dimension.

To proceed to the interpretation of Principal Components (PC's), we used Oblimin rotation, the same method used by [2]. We found the first PC is related to Reliability and the fourth PC is Tangibility. This two PCs are the same dimensions defended by the authors. The other PC's do not coincide exactly with the model; the second PC corresponds to Reputation, and is constituted by the questions G16, G18, G20, G21 and G22; third dimension can be assigned for Professors, and the last by Attendance Level (G19) - (Appendix 1). A measuring instrument such SERVQUAL scale needs to have trustworthiness and validity in acceptable degrees. To measure the reliability of the dimensions, we calculated Croanbach's Alpha coefficient to analyse the internal consistency of the scale for the variables (items) restrained in each dimension. We consider acceptable coefficients bigger or equal to 0.7 [12]. The values obtained for the Alpha of Croanbach in the 22 items was 0.86 (Appendix 2). That value indicates that items can be applied to the analysis with acceptable reliability.

4 Results and Discussion

The results of the questionnaire allow us to observe that the students who answered the survey attributed greater importance to the "capacity of the teaching staff" and the least important dimension was the "appearance of the physical equipment". In literature, we always find reliability first and tangibility last. The other three dimensions in the path without specific order a time that varies according to the area of application, therefore it coincides with our results. As followed by the authors, the GAP variable was calculated (Gap results from the difference between expectations and perceptions, and which will be identified by G1, G2, ..., G22). Then, principal component analysis (PCA) was applied. The values obtained for expectations are very high, resulting in negative gaps. Only the Tangibility dimension items show positive differences, so we can conclude that ESCE students are quite demanding and have high expectations. The data were adjusted for the application of the PCA using the Bartlett Test, where p-value = 0.000, and Kaiser-Meyer-Olkin Statistics (KMO) = 0.840. We used Kaiser criteria and scree plot to extract the five main components. The retained components explain 64.38% of the accumulated variance. To proceed with the interpretation of the components, oblique rotation was applied, as in the original SERVQUAL Model.

The use of expectations in measuring the quality of services is particularly important because they provide essential information in situations where different groups have similar perception values, which would make us think that there were no differences in the assessment of quality. However, analysing the expectations and consequently the values of the gaps (differences), the conclusions may change.

It is strategic for Higher Education Institutions to know the level of student satisfaction, since they act as disseminators of information by sharing their experience through forums, chats or the so-called "word of mouth", now supported by social networks that make it exponential [8]. A student's bad experience can take on overwhelming proportions if shared on social media.

Regarding the research questions, we can conclude that:

1. After applying the PCA, we had five dimensions according to the studies carried out by the authors [1, 2, 3]. However, the dimensions are not exactly the same. Although some of the dimensions are the same, there are differences that can be explained by the unique and specific characteristics of the HEI and by adapting the questions from the original questionnaire.

2. Dimensions do not all have the same relative importance.

3. From data analyse, it was possible to observe that the dimensions to which the respondents attributed greater importance was the "Reliability" and "Professors". These two dimensions are included as priorities in the strategic planning of ESCE – IPS. Reliability is measured and monitored through student surveys, control of the success rate in the course in terms of grades and student results. "Professors" dimension, it has been a constant commitment: the offer of pedagogical training, and in innovative pedagogical tools; support for scientific production and scientific projects, Erasmus mobility and support for participation in congresses and conferences. Collaborative work and the establishment of internal research networks were also encouraged.

The appearance of the physical equipment was the aspect valued as least important by the students. We can also observe that the values obtained for expectations were very high, which resulted in negative gaps. We can conclude that ESCE students are quite demanding, have high expectations or are very demanding in classifying perceptions, assigning them low values.

4. The dimension considered most important by the students who responded to the survey was the quality of teaching and the reputation of the institution (with 42.7%), followed by teaching capacity. The least important dimension was the appearance of the physical equipment (tangibility).

5 Final considerations

This study intended to draw attention to the problem of measuring students' expectations versus perceptions, which translates into a somewhat thankless task since the object of study is immaterial, and therefore there is an enormous subjectivity associated with this measurement.

In terms of practical implications, this study provides a structure and an application of SERVQUAL instrument that can be used by higher education institutions with a view to continuously improve educational quality [13].

Evaluating service quality level and understanding how various dimensions impact overall service quality enable educational institutions to efficiently design the service to provide a better allocation of resources and a better service to students [7]. Higher Education Institutions should concentrate their efforts on the perceived dimensions to focus their energy on a few attributes which are more important to service quality.

According to [17], whose article sought to gather articles on measuring the quality of care in HEIs through the HEdPERF scale. They found the HEdPERF scale application in studies in Brazil, China, Croatia, India, Malaysia, Portugal, Sri Lanka and Turkey [17, 18, 19].

5.1 Limitations and future research directions

Despite the consensus regarding the relevance of service quality in HEI, it is still a challenge to identify the most adequate measurement instrument [20] that allows best understanding of prior aspects and consequences of service quality, in order to promote methods that can improve quality and achieve a competitive edge.

When measuring customer expectations in terms of service quality, all research methods have limitations and weaknesses, so mixed models (qualitative and quantitative) should be used to minimize weaknesses and identify consistent information.

In this study, the SERVQUAL instrument is used to assess student satisfaction, through the difference between their expectations and perceptions [1, 2]. It should be noted that the results may have been influenced by the Covid 19 pandemic [15], since some of the students joined ESCE when we were in distance learning carried out through different platforms (zoom, moodle and Teams). Therefore, we think that this study would have better results if it also considered these other components or if it considered only perceptions, as there are critics who argue that expectations may be related to an ideal situation, an excellent institution, which does not exist. This study especially focused on SERVQUAL scales, which is no replacement for more specific research, considering other methods that may be applied in the educational sector for evaluating service quality such us HEdPERF or SERVPERF.

The data resulting from the SERVQUAL application must be analysed in detail in order to identify deficiencies in quality and define corrective actions.

Furthermore, when an HEI intends to implement quality improvement, it can use the SERVQUAL instrument, but with some restrictions [18], so that in future research other models already identified, HEdPERF [7, 11, 17] or EFQM [8] will be used.

References

1. A. Parasuraman, L. Berry, V. Zeithaml, Refinement and reassessment of the SERVQUAL scale, *Journal of Retailing,* **67**(4), 57-67 (1991)
2. A. Parasuraman, V. Zeithaml, L. Berry, A conceptual model of service quality and its implications for future research, *Journal of Marketing*, **49**, 41-50 (1985)
3. A. Parasuraman, V. Zeithaml, L. Berry, SERVQUAL: A multiple-item scale for measuring customer perceptions of service quality, *Journal of Retailing,* **64**, 12-40 (1988)
4. C. Gronroos, *Service management and marketing: A customer relationship management approach*, Second Ed., John Wiley & Sons (2000)
5. K. Albrecht, *Serviços com qualidade: A vantagem competitiva,* São Paulo: Makron Books (1992)
6. J. Quinn, Intelligent enterprise: *A Knowledge and Service Based Paradigm for Industry*, Free Press (1992)
7. F Abdullah, The development of HEdPERF: a new measuring instrument of service quality for the higher education sector, *International Journal of Consumer Studies*, **30**(6), 569 – 581 (2006), https://doi.org/10.1111/j.1470-6431.2005.00480.x

8. R. Laurett, L. Mendes, EFQM model's application in the context of higher education: A systematic review of the literature and agenda for future research, *International Journal of Quality & Reliability Management*, **36**(2), 257-285. https://doi.org/10.1108/IJQRM-12-2017-0282 (2019)
9. S. Hajdari, Service quality in higher education institutions an overview of models assessing it, *European Journal of Business and Management Research*, **4**(3), 1-4 (2019)
10. F. Abdullah, Measuring service quality in higher education: three instruments compared, *International Journal of Research & Method in Education*, **29**(1), 71-89, https://doi.org/10.1080/01406720500537445 (2006)
11. F. Abdullah, The development of HEdPERF: a new measuring instrument of service quality for the higher education sector, *International Journal of Consumer Studies*, **30**, 569-581 (2006)
12. M. Bujang, E. Omar, N. Baharum, A review on sample size determination for Cronbach's Alpha test: A simple guide for researchers. *Malays Journal of Medicine Science*. **25**(6), 85-99 (2018)
13. V. Teeroovengadum, T. Kamalanabhan, A. Seebaluck, Measuring service quality in higher education: Development of a hierarchical model (HESQUAL), *Quality Assurance in Education*, **24**(2), 244-258 (2016) https://doi.org/10.1108/QAE-06-2014-0028
14. P. Kotler, *Marketing Management – Global Edition*, Pearson Education, Inc. 16th ed. (2021)
15. M. Camilleri, Evaluating service quality and performance of higher education institutions: a systematic review and a post-COVID-19 outlook, *International Journal of Quality and Service Sciences*, **13**(2), 268-281 (2021) https://doi.org/10.1108/IJQSS-03-2020-0034
16. J. Abbas, HEISQUAL: A modern approach to measure service quality in higher education institutions, *Studies in Educational Evaluation*, **67**, 100933 (2020) https://doi.org/10.1016/j.stueduc.2020.100933
17. D. Silva, G. Moraes, I. Makiya, F. Cesar, Measurement of perceived service quality in higher education institutions: A review of HEdPERF scale use, *Quality Assurance in Education*, **25**(4), 415-439. https://doi.org/10.1108/QAE-10-2016-0058 (2017)
18. K. Fuchs, K. Fangpong, Using the SERVQUAL framework to examine the service quality in higher education in Thailand. *In: Education Quarterly Reviews*, **4**(2), 363-370 (2021)
19. F. Khattab, Developing a service quality model for private higher education institutions in Lebanon, *J. Mgt. Mkt. Review,* **3**(1), 24–33 (2018), https://ssrn.com/abstract=3157742
20. A. Brochado, Comparing alternative instruments to measure service quality in higher education, *Quality Assurance in Education*, **17**(2), 174-190 (2009)

Appendix 1

Questionnaire to Measure the Quality of the Services in a Higher Educations Institution (HEI) – PART A

Please, think about your concept of the ideal High School and the quality of services they provide. This questionnaire has two parts. The first part is concerned with your perceptions of Higher Education Institutions in general, while the second part is concerned with College of Business Administration (ESCE) – Polytechnic of Setúbal in particular.
Please, show the extent to which you think that HEI, in general, should possess the feature, please circle 7. If you strongly disagree that HEI should possess the feature, please circle 1. If your views are not strong circle one of the numbers in the middle. There are no right or wrong answers. All we are interest in are your expectations about Higher Education Institutions services.

1	2	3	4	5	6	7
Strongly Disagree	Disagree	Slightly Disagree	Neutral	Slightly Agree	Agree	Strongly Agree

1. Excellent HEI has modern technical equipment for the education process (i.e., computers/beamers)	
2. Physical facilities of excellent HEI are visually appealing (building and surroundings)	
3. Excellent HEI has physical facilities (parking, computers) in appropriate dimension to the number of students	
4. Teaching materials are available and up to date (study programs, brochures, student guides)	
5. Classes are held in accordance with the schedule of lectures and without delays	
6. When you have a problem, excellent HEI staff show a sincere interest in solving it	
7. The Excellent HEI takes care about the interest of the students (with the social aspects: residences, scholarships...)	
8. In Excellent HEI the plans of the courses and the programs are fulfilled in integrates	
9. The lessons given in excellent HEI are well prepared	
10. Academic staff of excellent HEI have scientific abilities to answer student's questions and doubts	
11. Academic staff of excellent HEI are always willing to help students (in classroom)	
12. Academic staff are available for consultations and are forthcoming with students	
13. Academic staff of excellent HEI have research productivity (publish books, magazine articles, doctoral thesis)	
14. Academic staff of excellent HEI have good communication skills	
15. Students of excellent HEI are informed promptly of the important dates such as: examinations, registrations, seminars	
16. Employers trust in the quality of education of excellent HEI	
17. Academic staff of excellent HEI apply pedagogic methods in their lessons	
18. In excellent HEI, the services supplied by sub-contracted companies (security, cleanness, bar) have good quality	
19. Excellent HEI have class time and office hours convenient to all their students	
20. The compositions of academic staff influence excellent HEI reputation	
21. The reputation of the HEI influences the choice of the employers	
22. The reputation of the HEI influence student's choice	

The 5 factors that characterize service quality are listed below. We ask you to distribute 100 points for each factor, according to the weight you consider they have in measuring the quality of services.

1. Appearance of physical facilities and communication materials ________
2. HEI ability to answer to student's necessities ________
3. Skills of the academic and non-academic staff ________
4. Confidence of academic and non-academic staff ________
5. Quality of educations and prestige of the Institution ________

TOTAL OF POINTS: **100 Points**

Questionnaire to Measure the Quality of the Services in a Higher Educations Institution (HEI) – PART B

The following questions are related to your feelings about ESCE. Please choose the number of the Likert-scale, that better represents your perceptions about ESCE.

1	2	3	4	5	6	7
Strongly Disagree	Disagree	Slightly Disagree	Neutral	Slightly Agree	Agree	Strongly Agree

1. ESCE has modern technical equipment for the education process (i.e., computers/beamers)	
2. ESCE physical facilities are visually appealing (building and surroundings)	
3. ESCE has physical facilities (parking, computers) in appropriate dimension to the number of students	
4. At ESCE, teaching materials are available and up to date (study programs, brochures, student guides)	
5. At ESCE, classes are held in accordance with the schedule of lectures and without delays	
6. When you have a problem, ESCE staff show a sincere interest in solving it	
7. ESCE takes care about the interest of the students (with the social aspects: residences, scholarships…)	
8. At ESCE the plans of the courses and the programs are fulfilled in integrates	
9. The lessons given in ESCE are well prepared	
10. ESCE academic staff have scientific abilities to answer student's questions and doubts	
11. ESCE academic staff are always willing to help students (in classroom)	
12. ESCE academic staff are available for consultations and are forthcoming with students	
13. ESCE academic staff have research productivity (publish books, magazine articles, doctoral thesis)	
14. ESCE academic staff have good communication skills	
15. Students of ESCE are informed promptly of the important dates such as: examinations, registrations, seminars	
16. Employers trust in the quality of education of ESCE	
17. ESCE academic staff apply pedagogic methods in their lessons	
18. At ESCE, the services supplied by sub-contracted companies (security, cleanness, bar) have good quality	
19. ESCE have class time and office hours convenient to all their students	
20. The compositions of academic staff influence ESCE reputation	
21. The reputation of the ESCE influences the choice of the employers	
22. The reputation of the ESCE influence students' choice	

Will you recommend ESCE to your friends and family? Yes ☐ No ☐

Personal Information:

Course: CF ☐ GRH ☐ MKT ☐ GDL ☐ GSI ☐

Year: 1º ☐ 2º ☐ 3º ☐

Gender: Male: ☐ Female: ☐

Age: 20 or less ☐ 21 – 25 ☐ 26 – 30 ☐ 31 – 35 ☐ more than 36

Situation: Full-time student ☐ Part-time student ☐ ☐

Thank you for your collaboration!!!

Appendix 2

Cronbach's Coeficients to the items and dimensions

Item	Dimensions - OBLIMIN	PC Loadings	Cronbach's Reliability coeficient (when item was deleted)
	RELIABILITY (0,845)		
G5	At ESCE classes are held in accordance with the schedule of lectures and without delays	0,857	0,805
G6	When you have a problem ESCE staff show a sincere interest in solving it	0,707	0,807
G7	ESCE take care about the interest if the students (with the social aspects: residences, scholarships,...)	0,75	0,814
G8	At ESCE the plans of the courses and the programs are fulfilled in integrates	0,818	0,806
G9	The lessons given at ESCE are well prepared	0,459	0,834
	REPUTATION (0,790)		
G16	The employers trust in the quality of education of ESCE	0,662	0,744
G18	At ESCE, the services supplied by sub-contracted companies (security, cleanness, bar,...) have good quality	0,726	0,734
G20	The composition of academic staff influences ESCE reputation	0,532	0,799
G21	The reputation of ESCE influences the choice of the entities employers	0,783	0,744
G22	The reputation of ESCE influences students choice	0,865	0,725
	PROFESSORS (0,907)		
G10	Academic staff of ESCE have scientific abilities to answer student's questions and doubts	-0,482	0,928
G11	Academic staff of ESCE are always willing to help students (in the classroom)	-0,886	0,879
G12	Academic staff are available for consultations and are forthcoming with students	-0,8	0,879
G13	Academic staff of ESCE have research productivity (publish books, magazine articles, doctoral thesis)	-0,829	0,88
G14	Academic staff of ESCE have good communication skills	-0,826	0,0884
G17	Academic staff of ESCE apply pedagogic methods in their lessons	-0,828	0,894
	TANGIBLES (0,685)		
G1	The faculty has modern technical equipment for the education process (i.e., computers/beamers)	0,482	0,694
G2	Physical facilities are visually appealing (building and surrondings)	0,665	0,606
G3	Physical facilities (parking, computers) in apropriate dimension to the number of students	0,763	0,619
G4	Teaching materials are available and up-to-date (study programs, brochures, student guides)	0,888	0,539
	ATTENDANCE LEVEL (-)		
G15	ESCE students are informed promply of the important dates such as: examinations, registrations, seminars	0,352	a
G19	ESCE has class time and office hours convinient to all their students	0,768	a

Using artificial intelligence as business opportunities on the market: An overview

Jakub Horák[1,*] and *Martina* Turková[1]

[1]Institute of Technology and Business in České Budějovice, School of Expertness and Valuation, Okruzni 517/10, 37001 Ceske Budejovice, Czech Republic

Abstract. The usage of artificial intelligence as business opportunities on the markets has a wide potential for many fields. Due to the rapid development of technology, artificial intelligence is quite expensive but useful. The research is done by making basic SWOT analysis (analysis Strenghts-Weaknesses-Opportunities-Threats) of the R&D (Research and Development) stage in the Czech Republic and by document analysis from articles on Web of Science and Scopus. The results shows, that the AI is beneficial for large number of fields and that the AI (Artificial Intelligence) has a great potential for implicating this in business, healthcare, economics and many more. The AI has potential to improve people´s life and work.
Keywords: Artificial intelligence, robotics, business opportunities, manufacturing company, finance

1 Introduction

Artificial intelligence (AI) is very extensive nowadays and many people do not realize what all falls under this term, even though they use it every day. [1] said that Artificial Intelligence has been rising as a powerful general-purpose technology that promises to lower uncertainty with more accurate predictions and to reduce the cost of laborious tasks. This paper focus on the issue of artificial intelligence and its implementation in business. This research informs the reader about the concept of artificial intelligence and what all falls into it. How it has developed and what are the benefits of its use, opportunities or threats. With the constant development of artificial intelligence, the properties and extent of application, as well as awareness of its existence, evolve. Artificial intelligence can be used in variety of fields, for example in customer service. As a primary AI tool, chatbots have seen increasing use by companies to support customer service. The difference between human intelligence and artificial intelligence is topic of many research, the, one of them was done by [2], in which he said that, these kinds of differences in basic structure, speed, connectivity, updatability, scalability, and energy consumption will necessarily also lead to different qualities and limitations between human and artificial intelligence. The co-working between those two are being more and more implicated in the workplace. In his research he discovered that humans working with AI systems in the workplace or in policy making have to develop an adequate mental model of the underlying `psychological' mechanisms.

* Corresponding author: horak@mail.vstecb.cz

However, the basic concept of artificial intelligence was to create new machines that were able to think as humans. Artificial intelligence helps to streamline activities that are routine in the process. If entrepreneurs, and not only them, understand the principle of functioning and improving performance using artificial intelligence, incorporation will gain positive results of society indicators. Artificial intelligence quickly moves into this world of expertise step by step by replacing people with higher predictive ability, higher efficiency and better results. However, homo sapiens will not be replaced by all at once, but gradually over time due to the absorption of systemic knowledge. Robotics become an essential part of our human body and existence.

[3] say that because of artificial intelligence, people in developed countries can worry about losing their jobs and, conversely, low -income countries can perceive artificial intelligence as an offer of new opportunities to break the poverty circle. Czechs are a bit sceptical for these technologies. A smaller part thinks that robotics are beneficial and most are worried about the loss of jobs, even though artificial intelligence is one of the tools to make work easier and more efficient. Basically, we already use it as a tool for work at work.

People are concerned about losing their jobs - it is perceived negatively, but AI also has a share in healthcare, where it helps to treat patients. Artificial intelligence is one of the areas of computer science with a mathematical process that has the potential to improve the health care system through new strategies to provide informed decision -making and facilitate patient involvement. It can be used to automate decision -making and prediction based on patient data. The characteristics of this technological progress are already influenced by every aspect of society and create conditions for the disruption of human socio -economic, educational, health, legal and moral system, which may have a significant impact on human progress [4].

The topic of business opportunities of artificial intelligence in the Internet environment will address questions such as their advantages and disadvantages in implementation. At the same time, there will be a notification on out possible pitfalls in their use, such as abuse in breaking protection or reducing the number of jobs on the market. Furthermore, the research will be focused on the use of artificial intelligence across various fields on the market and compare or evaluate their use and benefits.

2 Literature research

Artificial intelligence (AI), a key technology in the 2010s, has been increasingly becoming a dominant technology in the 2020 [5]. Artificial intelligence technology (AI) in the last decade, partially supported by high progress in the field of large data and raw computing performance, has entered the extraordinary phase of rapid development and wide application. Which corresponds with the research od [6], in which he said that, Artificial Intelligence is a disruptive technology developed during the 20th century, which has undergone an accelerated evolution, underpinning solutions to complex problems in the business world. Techniques developed in traditional AI research areas such as computer vision, speech recognition, natural language and robotics, have found many innovative applications in a number of real environments, including medicine. General methodological contributions of artificial intelligence, such as various recently developed deep algorithms, were also applied to a wide range of disciplines [7]. In other words, [8] said that Artificial intelligence (AI), particularly computational intelligence and machine learning methods and algorithms, has been naturally applied in the development of recommender systems to improve prediction accuracy and solve data sparsity and cold start problems.

[9] claim that artificial intelligence is known for its ability to let the machines perform tasks associated with the human mind and has useful applications in many industries and fields, including medicine and healthcare. Medicine continues to evolve because artificial intelligence, machine learning and deep learning in technology become more accepted and used. The research of [10] was focused on the trust management for AI, in particular, the importance of trust analysis and

management technology for artificial intelligence systems is continuously growing so that users who desire to apply and use artificial intelligence systems can predict and safely use services. The usage of the AI is become more and more involved in variety of fields. It's widely used in healthcare, economic and management, marketing and many more. One of them is hotel industry. The research on this topic was dome by [11]. The current study integrates these two concepts and explores how emotional and artificial intelligence influences employee retention and performance with a focus on service employees in the hotel industry. But the implementation of AI in services has some danger. As I said there are many differences between the human jobs and jobs done by the AI. The development of artificial intelligence (AI) does influence human jobs but not necessarily in a negative way. Although labor force participation rates and firms' job vacancies for human labor decline, the unemployment rate may be lower than that in an economy without AI [12].

2.1 Artificial Intelligence in economics

The importance of AI in economics is no doubt. Artificial intelligence can affect economic growth and employment. The influence is assumed to be substantial because the adoption of AI technology may lead to increased productivity, lower wages, prices, and labor substitution [13]. On the other hand [14] done research on modelling AI in economics, in their research they said, that it is also illustrate that the contribution of AI to aggregate productive labor service depends not only on the amount of AI services available but on the endogenous number of automated tasks, the relative productivity of standard and IT-related labor, and the substitutability of tasks.

2.2 Artificial Intelligence in healthcare

How can artificial intelligence improve existing approaches to supervision and response to public health and allow new ones? The answer lies in the basic challenges facing the supervision and response of public health. The supervision of public health is by its very essence controlled by data. Identification of early, accurate and reliable health anomalies and the outbreak of diseases from heterogeneous data sources have always been the main goal of supervision of public health. Artificial intelligence is one of the most exciting methodological shifts in our era. It holds the potential to transform healthcare as we know it, to a system where humans and machines work together to provide better treatment for our patients [15]. Technically, it is reflected in two different challenges: Promotions of data acquisition and analysis challenge. The first data acquisition prompt concerns the determination of easy -to -use data sources containing useful signals. The second analytical challenge concerns the development of efficient computational frames for extraction of such signals. Artificial intelligence provides a number of methods and techniques that help to deal with both challenges. Other main objectives and reactions of public health are to analyze and predict the trends of infectious diseases through modelling dynamics of disease transmission and evaluate public health reactions. Achieving these goals requires knowledge of domain and predictions rich in context, gentle risk analysis through contacts and social networks, quantification of reactions and control measures, and assessing these reactions and measures in the presence of complex interactions and restrictions.

Artificial intelligence methods can be very useful in building a virus detection system when there are too many data that people cannot easily handle. One example is usage of the AI during pandemic of COVID-19. During the pandemic, the researchers tried to predict patients trajectories to prevent more contacts, and become more aware of the spread.

Scanning system used at the airport for continuous scanning more faces. When AI methods are trained with some data on the properties of viruses, they can identify people who are patients without fever. Many researchers and scientists work on various methods such as Ann, machine learning and deep learning that can analyze available large data to detect signs of diseases before one can detect them with greater accuracy and efficiency [16].

One of the most important fields is neuroscience. Especially neurosurgery. Research done by [17] shows, AI is playing a pivotal role in the production, processing and storage of clinical and experimental data. AI usage in neurosurgery can also reduce the costs associated with surgical care and provide high-quality healthcare to a broader population.

2.3 Artificial intelligence and deep learning

Technological advances in artificial intelligence and data science have been a large amount of technological progress in the last decade. Although research in artificial intelligence for various applications has been underway for decades, the current wave of hype in the field of artificial intelligence has different from the previous ones. The perfect combination of higher computer processing speed, larger data libraries for collecting data and the large AI talent fund enabled the rapid development of AI tools and technologies, including in healthcare [18].

In particular, the development of deep learning (DL) had an impact on how we look at AI today and is the reason for a large part of the recent excitement around AI applications. DL allows you to find corrections that were too complex to render using previous machine learning algorithms [3].

2.4 Artificial Intelligence and finance

Prediction of the stock market is an important financial topic that has attracted the attention of researchers for many years. It includes the assumption that previous publicly available information has a certain predictive relationship to future shares revenues. By successful predicting future prices. Banks implement AI to provide digital assistance and financial advice to clients, measure their financial standing etc. [19].

Shares can be generated considerable profits. However, the actual pricing itself is much more demanding than just looking for a price direction. The methods listed in this article provide models to predict the stock price with reasonable accuracy. In addition, artificial intelligence machines can expand human abilities by intensifying their cognitive forces and can expand their physical abilities [20]. The sales forecast deals with the predication of the company's sales over a period of time in the near future, usually in the next fiscal year. Getting effective sales forecasts can help with decision -making powers to calculate production and material costs and allow them to determine the selling price, manage strategic operations, etc.

2.5 Artificial Intelligence in marketing

In this case, we are talking about machine-human interaction. The interaction between man and machine (HMI) indicates different ways of interacting people and automated systems and communication through touches, gestures, voice and sensors. HMI -based HMI applications include both people helping machines and machines helping people [21].

[22] mention that artificial intelligence systems and people can coexist. People can focus on the feeling of tasks, while artificial intelligence can be used as a tool that allows people to make better decisions. The study by [23] found that AI is highly improving the marketing market. The current study looks at how marketing managers utilize machine learning and artificial intelligence to analyze large amounts of client data and develop effective marketing strategies.

3 Methodology and data

Given the difficulty of a topic where it is difficult to do research using interviews or questionnaires, we will choose the methodology of document analysis. The research will be a gathering of information from professional articles, studies and literature and case studies.

We also will be making basic SWOT analysis on the Czech Republic stage of the Research and development (R&D).

For this research we chose the method of document analysis. The documents that the research is based on are professional articles will be sought after by the keyword "artificial intelligence" and these articles will be narrowed to economics and finance, business and other areas in business. In addition, publications from reliable scientific resources of two WOS and Scopus platforms will be used. Entering the criteria to website of Science, the publication will not be older than 2015 with limited articles, we have 85,636 articles that appear a key term "artificial intelligence." Scopus then offers us 71,772 articles. Articles are filtered from 2015 to the present and the type of documents are research articles, case messages and data articles. The other resources that will be used are official documents from the Office of the Government of the Czech Republic [24], which deals with the development of artificial intelligence in the Czech Republic.

4 Results

As we can see in the table 1 the SWOT analysis of the stage of Czech Republic.

Table 1. SWOT analysis of R&D in the Czech Republic (the CR).

Strengths	**Weaknesses**
- Infrastructure of research and development (R&D) for AI is on a great level *- The results of R&D can be compared to global quality levels* *- R&D application head to the fields with the greatest impart of AI* *- The confederation of Industry of the CR has established a platform for AI* *- Purposeful R&D support in AI has been growing for a long time* *- About 40AI start-ups operating in the Czech republic corresponds to comparable EU countries*	*- Application R&D does not have groundbreaking innovations* *- Low involvement in international R&D* *- Application R&D does not cover the fields of administration and trade* *- Purpose support for the applied LWS lags behind the support of basic research* *- The CR is a small market for AI start-ups* *- Risk investment in the CR are insufficiently used for groundbreaking technologies and are significantly lower in volume than abroad*
Opportunities	**Threats**
- Pressing foreign markets with advanced AI technologies of newly emerging small and medium-sized enterprises is essential for the Czech economy and can significantly improve the competitiveness of the CR *- Wider involvement in international research and international research associations* *- Improvement of cooperation between the public and private sectors in R&D and in the area transfer of R&D results into practice*	*- Insufficient AI reference in strategic-conceptual documents, long implementation of strategies* *- Foreign research will invest more and faster in AI* *- Brain outflow from research into multinational companies and abroad under better working conditions* *- Risk investment in start-ups is not compared to foreign volumes, lack of ground breaking technologies in the CR*

Source: Authors.

Figure 1 displays annual public support and the total cost of AI research and development between 2007 and 2019 (before 2007 is r it does not provide complete annual financial information on projects). Data for 2018 and 2019 are planned (was determined according to the planned budgets of ongoing projects).

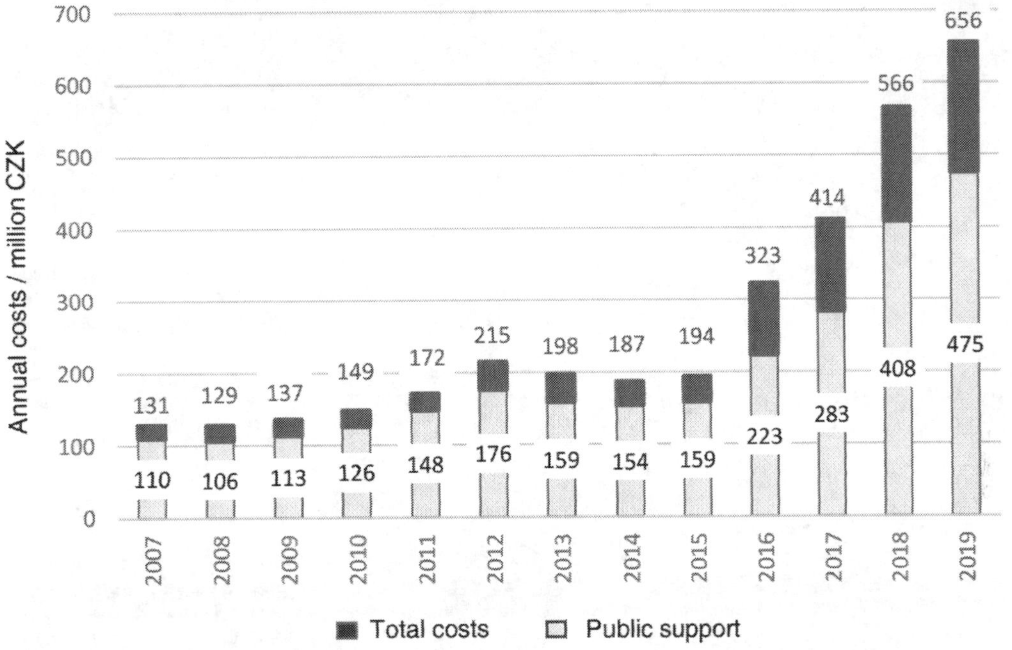

Fig. 1. Annual public support and total costs of research and development.

Source: [25].
Note: The number inside each column of the figure represents public support, the number always above the given column of the figure represents the total costs.

Many researchers have already illustrated the positive and negative impacts of the new network technology. 6G communication technology will be integrated with full artificial intelligence. In the 6G communication network, artificial intelligence will be an integral part of communication systems [26].

Figure 2 illustrates various new emerging technologies, complex and heterogeneous network structures, multi-spatial massive connectivity, and the wide range of available data across different structures and layers. An example of left horizons for providing high-performance continuous network connectivity is aerial Internet access. In the smart world of the future for travel, travelers across remote (oceanic) regions on ships and aircraft will also demand the same network service provision as is available to land/home users. To enable harmony across such massively interconnected complex 6G networks operating in coexistence with its predecessor will require tremendous learning and processing power [27].

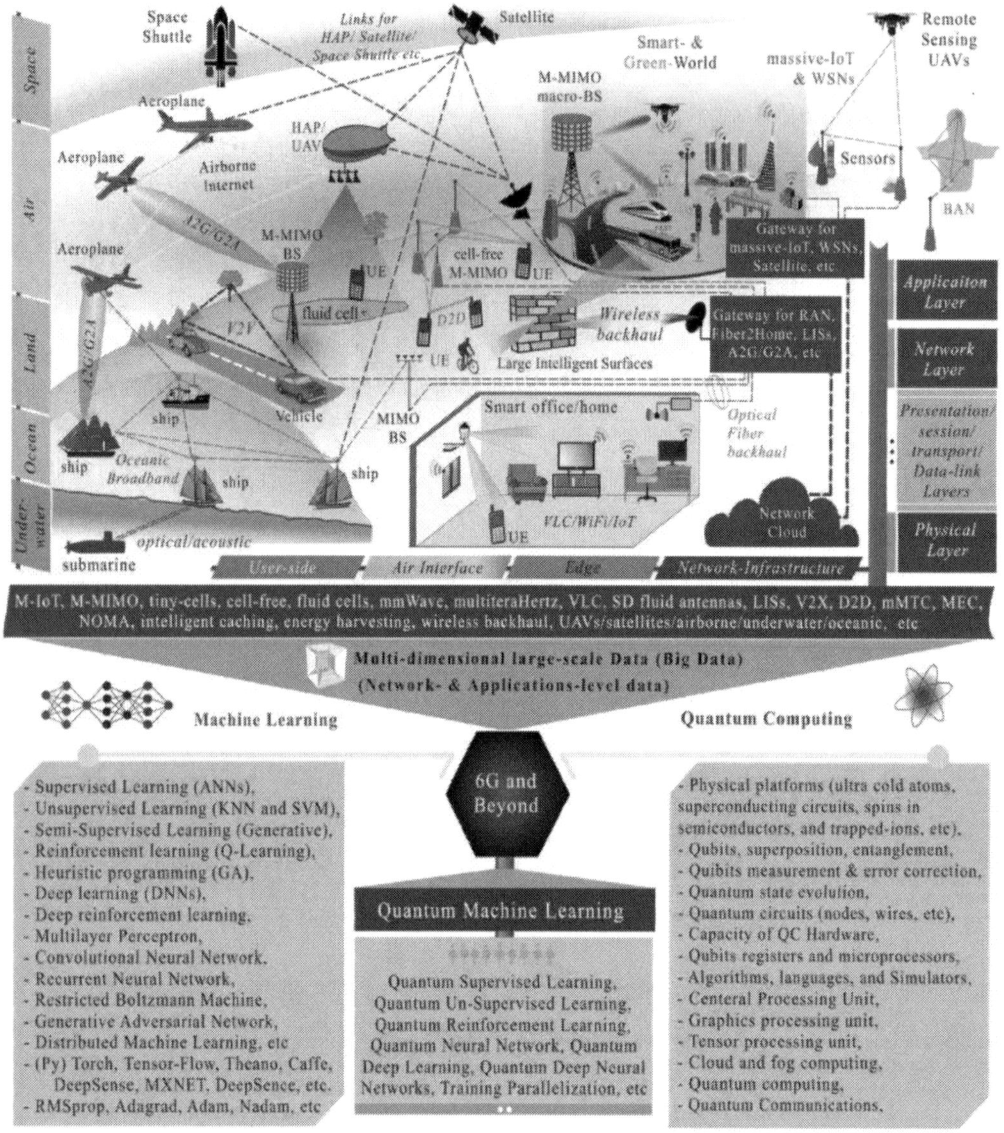

Fig. 2. Multi-Dimensional large-scale Data (Big Data).
Source: [27].

For some time on the market, the intelligence of artificial intelligence into households has been a great potential. Rocha et al. [28] claim that the Smart Home (SH) concept is part of a smart network paradigm that integrates smartness into the dwelling to achieve greater comfort with minimal possible energy expenditure. One of the problems to be solved is the management of demand (DSM), which consists of planning, introducing and monitoring the activities of the distribution network operator (bottom), the purpose of which is to allow more energy consumption to support efficient use of electricity.

The method of integrating artificial intelligence into a smart home aims not only user comfort but also energy consumption. Autors deal with numerical simulations with real data obtained from smart household appliances. The efficiency of the combination of artificial intelligence was reduced by 51% when comparing smart houses with and without and without a battery bank.

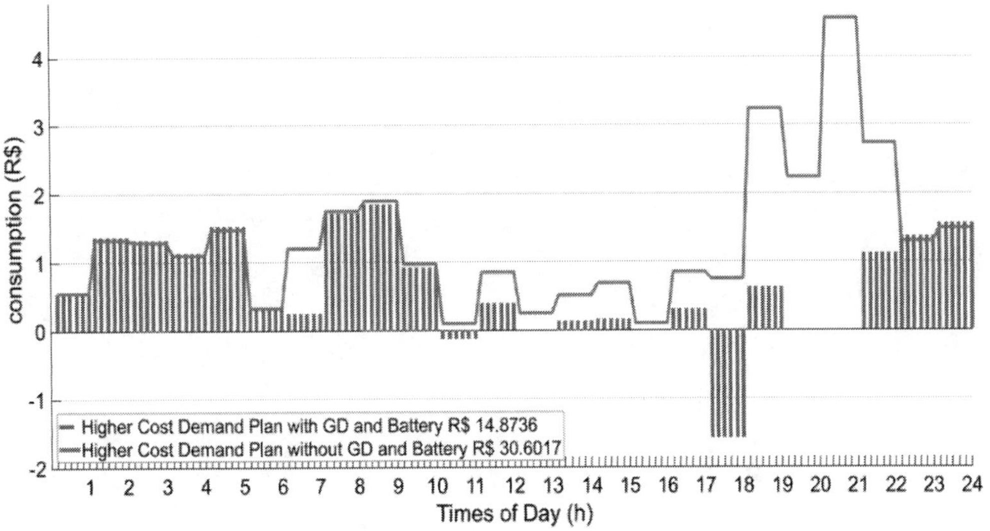

Fig. 3. Money consumption for the maximum cost of the maximum cost with and without GD and battery.

Source: [28].

Note: Solid line represents Higher cost demand plan without GD and battery, hatched line represents Higher cost demand plan with GD and battery.

The trend of today's time is the smart home or smart houses. Bicakci and Gunes [29] mention that nowadays homes with systems that can be accessed remotely to turn devices on and off are referred to as "smart". However, there is no intelligent automation process in such houses. Although such houses are considered smart, it is only because the word "Smart" is used as a marketing term. For example, Riverside Concept Houses declared that for a house to be defined as smart, it must have a system that controls electrical devices such as lighting, curtains - blinds, TV, air conditioning, sound system, etc. with a single remote control or via the Internet.

Table 2. Records of the PM2 Simulation and AI algorithm process.

App name	id	version	mode	pid	status	restart	uptime	cpu	men	user	watching
AI_choice_prediction	4	N/A	fork	1195	online	0	45D	0%	39,7 MB	root	disabled
AI_location_prediction	3	N/A	fork	1144	online	0	45D	0%	41,5 MB	root	disabled
simulation	1	N/A	fork	1037	online	0	45D	0%	39,3 MB	root	disabled
simulation_web_interface _control	2	N/A	fork	1068	online	0	45D	0%	206,7 MB	root	disabled

Source: [29].

https://doi.org/10.1051/shsconf/202316001012

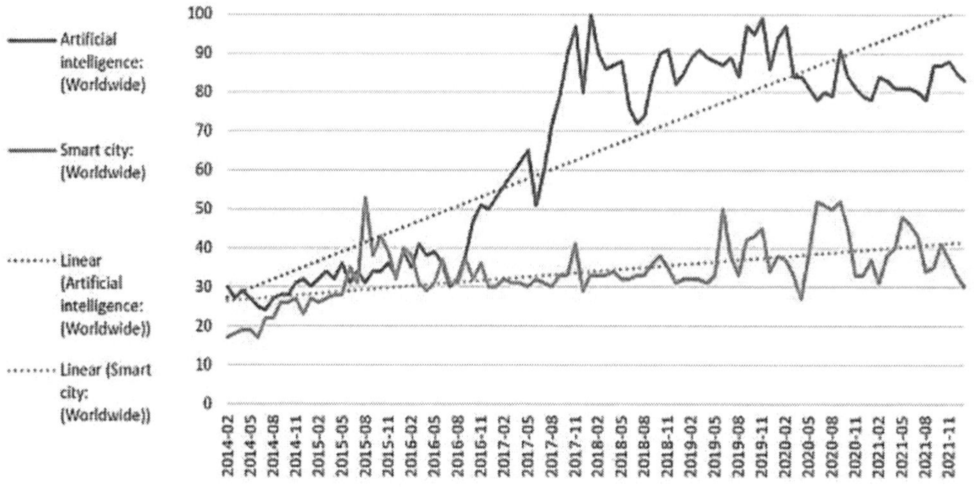

Fig. 4. Popularity of keywords "smart city and artificial intelligence".

Source: [30].

Note: The solid line peaking at 100 corresponds to Artificial intelligence: (Worldwide), the dotted line intersecting this curve represents Linear (Artificial intelligence: Worldwide), the solid line below represents Smart city: (Worldwide) and the dotted line intersecting this curve represents Linear (Smart cities: (Worldwide)).

As modern smart cities, we can imagine various uses of digital, information or communication technologies to reduce consumption, comfort of the inhabitants of these cities and streamline infrastructure.

Artificial intelligence enabling solutions for smart cities has several advantages, including more reasonable water supply, energy management, waste management and reducing traffic columns, noise and pollution. Most of the activities and technologies of a smart city focused on creating data and getting new information about the complexity of the city and dynamics [31].

5 Discussion

Artificial intelligence brings new opportunities and benefits to society. Figure 1 shows a gradual increase in the cost of the project of artificial intelligence. Research and development (R&D) of artificial intelligence is better than abroad. However, the involvement in international R&D is low. The R&D applications are heading to fields where artificial intelligence has the greatest impact. The Czech Republic is a small market for AI start-ups. In order to increase the competitiveness of the Czech Republic, the development and emergence of new start-ups of companies with high added value in AI technology is important.

Furthermore, according to the SWOT analysis between AI opportunities, we include the preparation of the national strategy for the implementation of artificial intelligence. Artificial intelligence with its technologies can capture newly emerging small and medium -sized enterprises on foreign markets and can improve the competitiveness of the Czech Republic. The threat to the Czech Republic is then greater and faster investment of foreign research and we will be lagging behind international research due to loss of cooperation with foreign leaders in the field.

In Figure 2, we can see the size of the range of artificial intelligence in the world. Artificial intelligence is in the form of virtual assistants, software and systems, for example, to recognize face or voice or in the form of robots, drones and internet of things. Nowadays, we can notice the extent of the use of artificial intelligence, for example, when searching on the Internet, shopping online, mobile phone assistants who help us organize and perform tasks, answer questions through a virtual assistant and help with search or recommendations. Furthermore, they are smart domestic products - vacuum cleaner, refrigerator, television, thermostats or even washing machines used to reduce consumption, autonomous vehicles and house security, as well as security on the Internet and fight disinformation. Artificial intelligence has a large share in health care and was also used in the fight against Covid.

As artificial intelligence makes its progress and is constantly evolving in something through products known as smart appliances, smart mobile phones or smart homes that grow in "smart cities". The aim is to improve people's quality of life, increase the efficiency and availability of public and social services. These elements are incorporated in our country, for example, in public transport - contactless terminal for payments, in Ostrava there are smart stops with digital information panels. Unfortunately, the unsuccessful example of the involvement of artificial intelligence was smart benches with solar panels or smart lamps. We think that is a matter of time before such errors with dysfunction and the elements will be integrated for the test in larger cities.

The benefit of the paper is to raise awareness of the existence and development of artificial intelligence in the world, to present fields where artificial intelligence is currently used and what it has. Artificial intelligence is a great benefit for many, for achieving companies and organizations more efficient, improving the quality of life and facilitating manual work. Artificial intelligence helps to investigate the environment, energy consumption, while checking the functionality of technology and their security. It also helps to safety of real estate and other property, but also the health of persons. This is the aim of the work to realize society.

6 Conclusion

The article deals with the current topic of the use of artificial intelligence in the world. In the theoretical part of the article, examples of artificial intelligence in various fields are given. We mentioned that the most use of AI is in healthcare. We think that those who do not work in healthcare do not even realize the extent to which AI is used here, and many end consumers perceive the use of artificial intelligence on smartphones or in homes. Due to the rapid development of artificial intelligence, the Internet of Things and overall digitization, it is necessary for us to constantly educate ourselves and know how to use AI effectively. It is determined that, to understand and accept AI (include it into decision-making processes and business practices), institutions are required to make technologies more understandable for perception.

There are also negative comments and concerns about this discussed topic, especially concerns about the loss of jobs and their replacement by robots. Robots are of course used for various types of activities. But it is still necessary to check the work after these robots, so the robots aim to make stereotypical and manual work more efficient.

In the second part of the paper we stated how the Czech Republic is doing in the field of AI research and development. The scope of the implementation of artificial intelligence is pointed out here, whether it is transport – air, water or land, smart home, smart city, but also marketing and healthcare. The concept of smart home, smart appliances that are supposed to save consumption is explained in more detail here. New family houses and apartments already count on the introduction of a smart home during construction and adapt them to it. Individual elements of smart cities are currently being tested in larger cities such as Prague, Brno or Ostrava. In the future, the technologies used will also absorb a number of smaller cities. For the young generation, such a step forward is a great benefit, because this generation is close to new technologies. The introduced technologies mainly have the goal of greater awareness of improving the quality of life and finding appreciation even by older generations.

By introducing AI technologies, the state is mainly inspired by foreign countries, where AI has more experience and development is many times faster here.

Acknowledgements

This research was supported by the project, which has received funding from the Institute of Technology and Business. Grant Number IVSUZO001 "Benchmark of the world economy in the context of the principles of the circular economy and determining the value of the company". This paper was also supported from the project of Specific University Research, reg. number 05SVV22 – Scientific School and its connection with the experts from practice in the research activities of the institution.

References

1. E. Kalinova, Artificial intelligence for cluster analysis: case study of transport companies in Czech Republic. *Journal of Risk and Financial Management*, **14**(9), (2021)
2. J. E. H. Korteling, G. C. Van De Boer-Visschedijk, R. A. M. Blankendaal, R. C. Boonekamp, A. R. Eikelboom, Human- versus artificial intelligence. *Frontiers in Artificial Intelligence*, **4** (2021)
3. Bohr, K. Memarzadeh, The rise of artificial intelligence in healthcare applications. *Artificial Intelligence in Healthcare,* 25-60 (2020)
4. R. S. Ravali, T. M. Vijayakumar, K. S. Lakshmi, D. Mavaluru, L. V. Reddy, Retnadhas, M., T. Thomas, A systematic review of Artificial intelligence for Pediatric Physiotherapy practice: Past, Present, and Future. *Neuroscience Informatics*, 100045 (2022)
5. T. Krulicky, E. Kalinova, J. Kucera, Machine learning prediction of USA export to PRC in context of mutual sanction. *Littera Scripta*, **13**(1), 83-101 (2020)
6. J. L. Ruiz-Real, J. Uribe-Toril, J. A. Torres, J. De Pablo, Artificial intelligence in business and economics research: Trends and future. *Journal of Business Economics and Management*, **22**(1), 98-117 (2021)
7. P. Suler, Z. Rowland, T. Krulicky, Evaluation of the accuracy of machine learning predictions of the Czech Republic´s exports to the China. *Journal of Risk and Financial Management*, **14**(2), (2021)
8. M. Vochozka, J. Horak, T. Krulicky, Innovations in management forecast: Time development of stock prices with neural networks. *Marketing and Management of Innovations*, **2020**(2), 324-339 (2020)
9. Y. Park, D. Casey, I. Joshi, J. Zhu, F. Cheng, Emergence of new disease: How can artificial intelligence help? *Trends in Molecular Medicine,* **26**(7), 627-629 (2020)
10. T. W. Um, J. Kim, S. Lim, G. M. Lee, Trust management for artificial intelligence: A standardization perspective. *Applied Sciences,* **12**(12) (2022)
11. C. Prentice, S. D. Lopes, X. Wang, Emotional intelligence or artificial intelligence– an employee perspective. *Journal of Hospitality Marketing & Management*, **29**(4), 377-403 (2020)
12. M. Vochozka, J. Horak, T. Krulicky, P. Pardal, Prediction future Brent oil price on global markets. *Acta Montanistica Slovaca*, **25**(3), 375-392 (2020)
13. L. Wang, P. Sarker, K. Alam, S. Sumon, Artificial intelligence and economic growth: A theoretical framework. *Scientific Annals of Economics and Business,* **68**(4), 421-443 (2021)

14. T. Gries, W. Naude, Modelling artificial intelligence in economics. *Journal for Labour Market Research,* **56**(1) (2022)
15. M. Pedersen, K. Verspoor, M. Jenkinson, M. Law, D. F. Abbott, G. D. Jackson, Artificial intelligence for clinical decision support in neurology. *Brain Communications,* **2**(2) (2020)
16. S. Agrebi, A. Larbi, Use of artificial intelligence in infectious diseases. *Artificial Intelligence in Precision Health,* 415-438 (2020)
17. M. Mofatteh, Neurosurgery and artificial intelligence. *AIMS Neuroscience*, **8**(4), 477-495 (2021)
18. T. W. Um, J. Kim, S. Lim, G. M. Lee, Trust management for artificial intelligence: A standardization perspective. *Applied Sciences,* **12**(12) (2022)
19. M. Ryzhkova, E. Soboleva, A. Sazonova, M. Chikov, A. Vankevich, T. Ilina, Consumers' Perception of Artificial Intelligence in Banking Sector. *SHS Web of Conferences,* **80** (2020)
20. P. K. Kopalle, M. Gangwar, A. Kaplan, D. Ramachandran, W. Reinartz, A. Rindfleisch, Examining artificial intelligence (AI) technologies in marketing via a global lens: Current trends and future research opportunities. *International Journal of Research in Marketing,* **39**(2), 522-540 (2022)
21. H. J. Wilson, P. Daugherty, Collaborative intelligence. *Harvard Business Review,* **96**(4), 114-123 (2018)
22. A. Kaplan, M. Haenlein, Siri, Siri, in my hand: Who´s the fairest in the land? On the interpretations, illustrations, and implications of artificial intelligence. *Business Horizons*, **62**, 15-25 (2019)
23. M. S. Hossain, M. F. Rahman, Machine learning and artificial intelligence: The new move for marketers. developing relationships, personalization, and data herald in Marketing 5.0. *IGI Global,* 215-241 (2022)
24. Office of the Government of the Czech Republic [online], Available at: https://www.vlada.cz/assets/evropske-zalezitosti/aktualne/AI-technologie-2018.pdf (2018)
25. Science and Research, Artificial intelligence research in the Czech Republic. What are its results and where does public support go? [online], Available at: https://vedavyzkum.cz/z-domova/technologicke-centrum-av-cr/vyzkum-umele-inteligence-v-cesku-jake-jsou-jeho-vysledky-a-kam-smeruje-verejna-podpora (2020)
26. B. McMahan, E. Moore, D. Ramage, S. Hampson, B. A. Arcas, Communication-efficient learning of deep networks from decentralized data. *Artificial Intelligence and Statistics*, 1273-1282 (2017)
27. S. J. Nawaz, S. K. Sharma, S. Wyne, M. N. Patwary, M. Asaduzzaman, Quantum machine learning for 6G communication networks: State-of-the-art and vision for the future. *IEEE Access*, **7**, 46317-46350 (2019)
28. H. R. Rocha, I. H. Honorato, R. Fiorotti, W. C. Celeste, L. J. Silvestre, J. A. Silva, An Artificial Intelligence based scheduling algorithm for demand-side energy management in Smart Homes. *Applied Energy,* **282**, 116145 (2021)
29. S. Bicakci, H. Gunes, Hybrid simulation system for testing artificial intelligence algorithms used in smart homes. *Simulation Modelling Practice and Theory,* **102** (2020)
30. H. M. K. K. M. B. Herath, M. Mittal, Adoption of artificial intelligence in smart cities: A comprehensive review. *International Journal of Information Management Data Insights*, **2**(1), (2022)
31. N. Kapoor, N. Ahmad, S. K. Nayak, S. P. Singh, P. V. Ilavarasan, P. Ramamoorthy, Identifying infrastructural gap areas for smart and sustainable tribal village development: A data science approach from India. *International Journal of Information Management Data Insights,* **1**(2) (2021)

Using social networks in business

Eva Kalinová[1,*] and *Hana* Kovaříková[1]

[1]Institute of Technology and Business, School of Expertness and Valuation, Okruzni 517/10, 37001 Ceske Budejovice, Czech Republic

Abstract. Social media platforms and channels are being increasingly used nowadays, influencing almost all types of businesses. Social media are mainly used as a source of information and means of dissemination of knowledge that may change the opinions of people. They can be considered very important business marketing strategies since they participate significantly in creating new business opportunities, building a stronger position in the market, or in the transformation of consumer behaviour. The main goal of the paper is to determine which social networks are most commonly used in business and their purpose. To achieve the goal, the method of questionnaire survey was used, in which a total of 237 respondents participated. The results of the survey show that business entities mostly use Instagram in their business activities, mainly for promoting their company, which was followed by Facebook, through which they sell their products or attract new employees. In addition, companies also use social networks as a communication channel with employees or customers.
Keywords: Communication, society, social media, customers, advertising

1 Introduction

Today´s time is characterized by various innovations and modern technologies, including social networks, which can be considered an integral part of everyday life [1]. In nearly all parts of the world, countless people, both young and adult, old or even retired use social networks on everyday basis. Today, practically every person has the opportunity to communicate via computer, notebook, tablet, smart phone, etc. with any person registered on a given social network [2].

Social media is a global phenomenon that is used daily by several billion users. Obada and Dabija [3], see social media as a means of communication and uniting people around the world, which enable people to learn about new customs and cultures and enhance mass dissemination of information among people (sharing videos and photos, job offers, reviews, etc.). Recently, they have been widely applied in the business activities of business entities. According to [4], the influence of social media enters all aspects of human life, both personal and professional. Business entities try to use the power of social media mainly for the promotion of their products, building their brand, or attracting new customers. Social networks help in building the loyalty of the existing customers to the point that they will

* Corresponding author: kalinova@mail.vstecb.cz

voluntarily and for free increase the awareness of the qualities of a given company. Possible users of social networks include both individuals and groups of employees or marketing managers themselves.

Pavlíček [5] states that the adoption of social networks by companies in recent years has enabled creating new and more interesting social networks. Any social network is based on common interests, family ties, or other more pragmatic reasons, such as economic, political, or cultural interests. The main reason for business entities to use social networks in their business activities is the extent of the field of their activities. An example can be Facebook with the number of users exceeding two billion.

There is no doubt that social networks in general open new challenges and opportunities but on the other hand, they may also pose threats and misused opportunities [6].

The main goal of the presented paper is to identify the social networks most used in business and their purpose.

To achieve the set goal, the following research questions are formulated:

RQ1: For what purposes do companies use social networks?

RQ2: What communication channels do companies use?

2 Literature research

The present turbulent development of information technologies forces companies to adapt to the current trends in their business activities, especially those concerning IT [7]. This applies even more to business entities that carry out their business activities in the online environment. One of the most used and pronounced current trends include social networks [8].

As Tiwasing [9], states, the term social networks/social media emerged quite recently in connection with the emergence of various technologies such as RSS and blogs and online resources (on-line social networks, video hosting, etc.). Although the above-mentioned technologies might seem not to have much in common, this is not entirely the case, as all of them serve to facilitate the exchange of information between users compared to the technology of the previous generation when the Internet consisted predominantly of static pages [10].

According to Jurišová [11], social media have several basic typical features. Specifically, the author states that social media stimulate users to communicate, respond or provide feedback for everyone connected. Social media can be seen as an open system with almost no barriers, providing two-way communication; support effective communication of communities [12].

Rakshit et al. [13], argue that social networks can be used by companies as a way of advertising, employee communication, or as an intermediary between a company and a customer or potential customer.

Social networks enable people to keep in touch with their friends, share their experiences, or even find work. Although social networks might appear to be at their top and not evolving further, it is not true. Social networks are still one of the fastest growing internet environments, oriented not only towards individuals and groups but also an effective tool for internet marketing [14].

The following text briefly describes some of the existing social networks.

One of these networks is e.g. the social platform Facebook. Facebook can be seen as a means to satisfy the human need for sharing, for having something in common, including everyday activities [15]. Facebook is not only a page for posting but also a platform, a place where it is possible to run various smaller applications [16]. Social network LinkedIn aims to create a professional network that would enable increasing the connectivity of contacts between individual employees and establish closer contact between them. LinkedIn was initially intended for finding contacts with people we need through people we already know and trust. According to [17], it enables creating a professional identity on the Internet and finding and obtaining job offers. LinkedIn provides numerous advantages that can be used for professional development and starting the desired career. It enables creating professional profile, uploading CV and the information about education and work experience and thus attracting potential employers and obtain the desired job [18]. Social network Twitter was created in 2006 and works on the basis of the so-called microblogs. Microblogs are short messages whose length does not exceed 140 characters. This network was primarily intended for travelling people so that they could quickly share their experiences through mobile phones or other devices [19].

Another very popular social network is Instagram, a free application for sharing photos and videos available for iPhones and Android devices. People can use Instagram to share photos and videos with their followers or selected groups of friends. There are also used so-called hashtags that enable easier search for the content with the same topics [20].

3 Data and methods

The basic data source was the responses to questions asked within the first questionnaire survey, which aimed at achieving the set goal of the paper presented. The effectiveness of the research results to a certain extent depends on a suitably chosen sample of interviewed respondents. The questionnaire was distributed to 247 social network users, and a total of 237 completed questionnaires was obtained.

The questionnaire was created using websites, specifically survio.com, and consisted of 8 questions. For convenience, the respondents could choose from the proposed options. The structure of the questionnaire is as follows: in the introductory part, the respondent is informed about the purpose of the questionnaire and its approximate length, with included request for the completion of the questionnaire by the respondent. The first three questions are general: the questions from the fourth one concern the issue of social networks.

The second data source is another questionnaire, which was created for an unnamed company. Based on this questionnaire, it was found which social networks are used by companies and for what purpose, whether they use social networks only within the company, whether they are available for their customers, and whether they use them to sell their products.

All answers will be processed graphically and then interpreted.

4 Results

The first question aimed to determine whether social networks are more used by men or women (see Figure 1).

As seen below, social networks are more used by women (133 women) compared to 104 men; the difference is thus not significant.

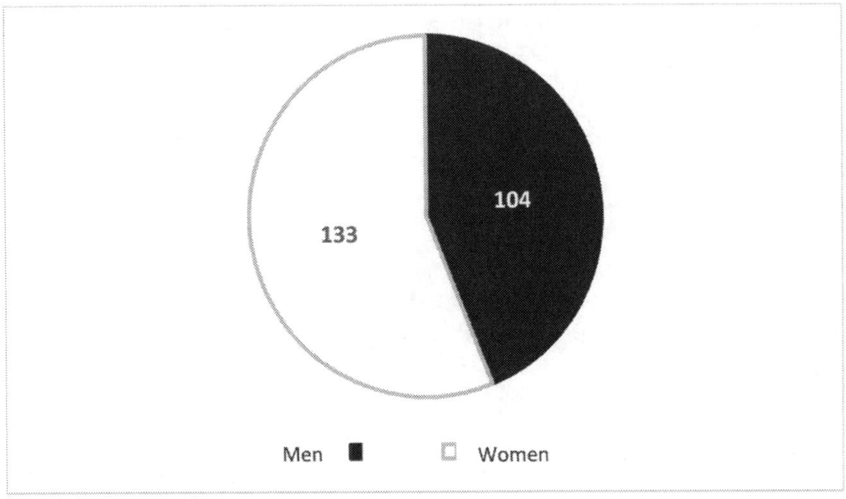

Fig. 1. Social network users.

Source: Own processing.

Furthermore, it was determined by which age groups are the social networks most used. The group aged 18-24 dominated the largest share of social network usage. The age group 17 and under and 25 - 34 were in the background. These groups were the focal point for gathering information and evaluating data. The 45-54 and 55-64 age groups do not use social networks so significantly to increase the rating of businesses. The last group of 65+ does not participate in the use of social networks almost at all.

The responses to this question are graphically processed in Figure 2.

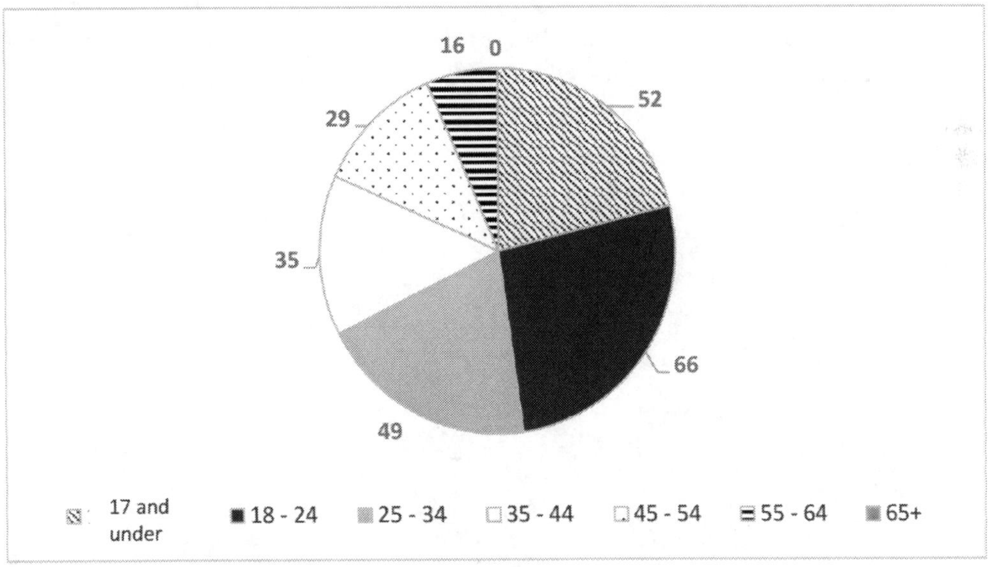

Fig. 2. Age structure of social networks users.

Source: Own processing.

The third question focused on the most used social networks. The responses can be seen in Figure 3.

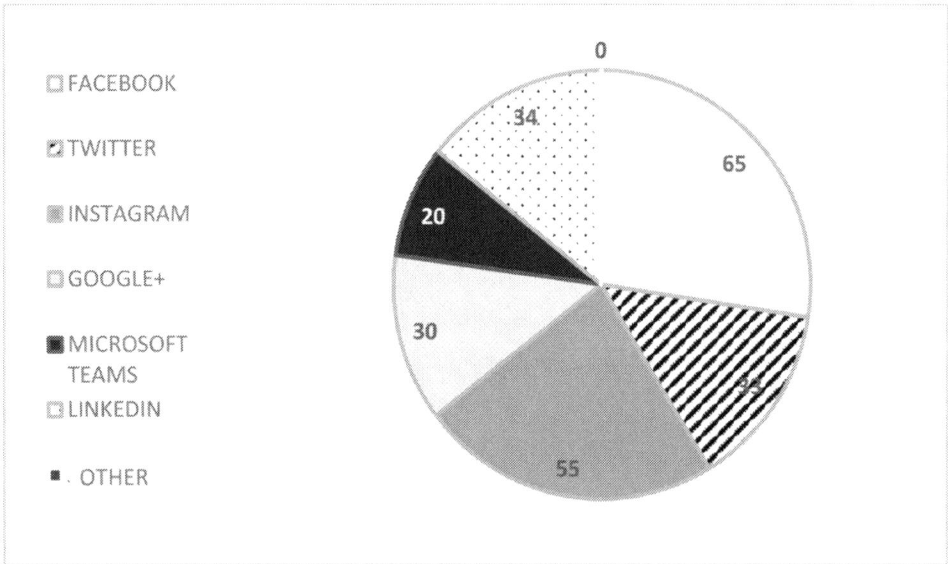

Fig. 3. Social networks used by the respondents.

Source: Own processing.

Figure 3 shows that the most commonly used social networks are Facebook (65 respondents) followed by Instagram (55 respondents) and Microsoft Teams (20 respondents).

The next part of the questionnaire focused on information concerning the issue obtained from an unnamed company, i.e., which social networks the company uses. The rest of the platforms are used less by the respondents, in which Linkedin, Google+ and others were evaluated. The responses are processed graphically in Figure 4.

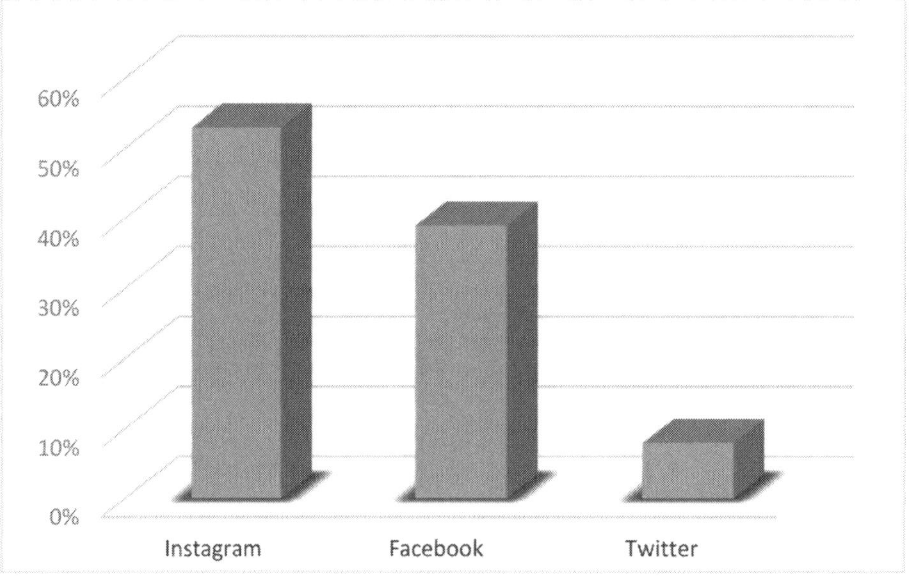

Fig. 4. Social networks used by the addressed company.

Source: Own processing.

As can be seen in the figure, the company mostly uses Instagram, Facebook, and sometimes Twitter, out of which Instagram represents the largest share. Instagram is used by the company both for promotion and to obtain feedback from customers, which means that Instagram enables the evaluation of its functioning. Facebook is used for recruiting new employees, posting vacancies, and also for selling its products. Twitter is the least used platform, used by companies for the purposes of posting microblogs but not very often.

Figure 5 below shows the purpose of using social networks as interpreted above.

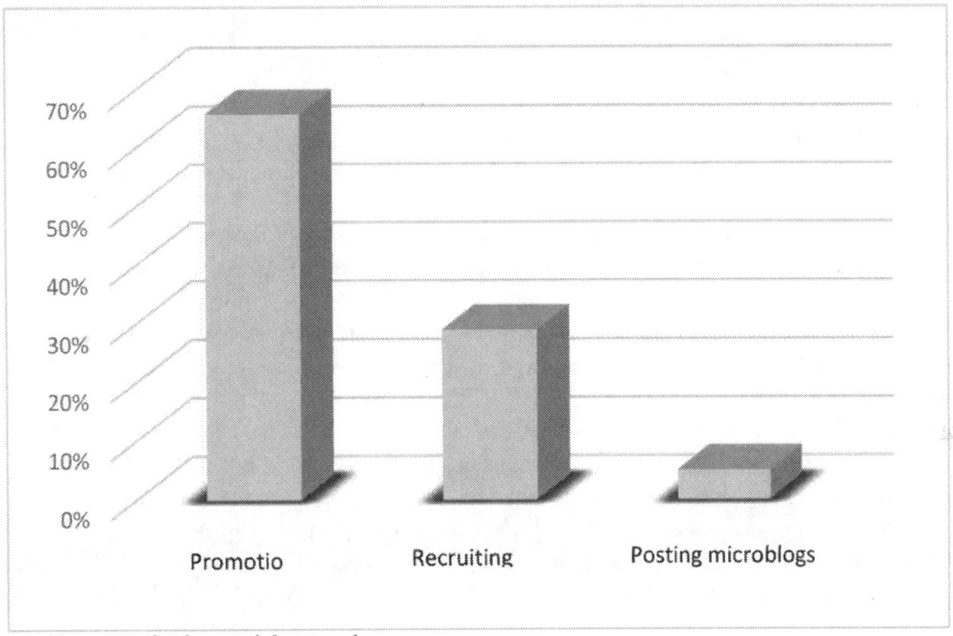

Fig. 5. Purpose of using social networks.

Source: Own processing.

5 Discussion of Results

Based on the obtained results, it is possible to answer the formulated research questions concerning for what purpose the company addressed uses social networks. The company uses social networks mostly for the promotion of the company in order to increase its competitiveness among other companies.

The research shows that the most widely used platform for this purpose is Instagram, where the company has its account and publishes various posts concerning new information about the company, successes, or who is interested in the company. It is also possible to obtain feedback from its customers, who can evaluate the company, add reviews and comments, follow the posts and share them with their friends. Furthermore, it was found that the company uses Facebook for selling its products or for recruiting new employees. It has the opposite effect when customers or potential employees share the posts concerning either vacancies or products being sold. The last platform used by the company is Twitter, which serves mainly for sharing short posts not exceeding 140 characters, which are also referred to as microblogs.

The communication channels used by the company include Instagram, Facebook, and Twitter. A business could focus on other social media platforms, but that would require much more extensive management and care. The most important point is communication with customers to create comfort and an active approach regarding their satisfaction. The use of social networks increases consumer perceptions of mutual involvement in customer purchases of goods or services. Customer reviews, discussion forums and product scores provide important data for rating overview. The collection of information further informs customer decision-making. The question remains of whether the company should focus on other applications besides the promotion of the company, either communication with customers or communication among employees, which is an important step for the prosperity of the company.

6 Conclusion

This post aims to provide answers to questions that the lay public usually cannot answer, i.e., what social networks companies use and for what purpose. In our article, it was found that social networks can be used to promote a company. Representations of the most used platforms according to the results were shown in the graph, where the highest user rank was occupied by Facebook and Instagram. The platforms Linkedin, Twitter, Google+, Microsoft Teams and others were placed in the background. The main reasons for companies using social networks were represented in a bar graph, where the most important activity of the companies was their promotion of the company. In other words, companies want to achieve the highest possible visibility on the competition market. Another reason was regarding the recruitment of new employees for agreeing certain dates for recruitment interviews. There are still a lot of people who are unaware of this fact or unfamiliar with the subject; even users of social networks often do not think to investigate this issue in more detail. This post may be of benefit to people who have never considered using social media for this purpose or to people who do not use social media at all. Based on the use of social networks, people can obtain a lot of information and opportunities, whether work or otherwise. Perhaps it would be appropriate to carry out further research, which would also cover the negative effects of social networks.

The aim of the contribution was to identify social networks used by companies and their purpose. The purpose of the post has been met. Data obtained from a questionnaire survey of distributed users of social networks and unnamed companies was analyzed and processed. The questionnaire was created through the website survio.com, where the options of possible answers were also obtained.

The obtained results were analyzed and evaluated in the Discussion. According to the analysis, the companies should improve communication to achieve even better results. At the end, the benefits of the submitted contribution and recommendations for possible further research were presented.

The main limitation of the research was that the approached companies were not willing to share internal information. It was therefore necessary to modify one of the questionnaires to obtain some information.

References

1. J. Hollowell, Z. Rowland, T. Kliestik, J. Kliestiková, V. Dengov, Customer loyalty in the sharing economy platforms: How digital personal reputation and feedback systems facilitate interaction and trust between strangers. *Journal of Self-Governance and Management Economics*, **7**(1),13-18 (2019)
2. E. Kalinová, A. Neubergová, Influencer communication on the social network Instagram. *AD ALTA: Journal of Interdisciplinary Research*, **11**(2), 107-111 (2021)
3. D. R. Obadă, D. C. Dabija, The mediation effects of social media usage and sharing fake news about companies. *Behavioral. Sciences,* **12**(10) (2022)

4. J. M., Balogh, T. Mizik, Impacts of marketing strategy and social media activity on the profitability of online wine shops: The case of Hungary. *Economies,* **10**(12) (2022)
5. A. Pavlíček, *Nová média a sociální sítě [New media and social networks].* 1st ed. Prague: Oeconomica. ISBN 978-80-245-1742-1 (2010)
6. G. Lăzăroiu, M. Kováčová, A. Siekelová, J. Vrbka, Addictive behavior of problematic smartphone users: The relationship between depression, anxiety, and stress. *Review of Contemporary Philosophy*, **19**, 50-56 (2020)
7. C. Luo, Analyzing the impact of social networks and social behavior on electronic business during COVID-19 pandemic. *Information Processing and Management*, **58**(5) (2021)
8. I. R. P. Cuypers, G. Ertug, J. Cantwell, A. Zaheer, M. Kilduff, Making connections: Social networks in international business. *Journal of International Business Studies,* **51**(5), 714-736 (2020)
9. P. Tiwasing, Social media business networks and SME performance: A rural-urban comparative analysis. *Growth and Change,* **52**(3),1892-1913 (2021)
10. E. Ahumada-Tello, R. Ravina-Ripoll, E. J. Galvez-Albarracin, Social networks and academic performance self-perception in business sciences students. Redes sociales y autopercepción de desempeño académico en estudiantes de ciencias empresariales. *Cuadernos de Administracion-Universidad Del Valle,* **36**(66),105-117 (2020)
11. E. Jurišová, The impact of social networking on business and business ethics. *Conference: Management Challenges in the 21st Century,* 403-411 (2013)
12. R. Bílková, Z. Zelenka, Social Network Marketing: An examination of marketing behavior of small businesses. *26th International-Business-Information-Management-Association Conference*, 2171-2180 (2015)
13. S. Rakshit, N. Islam, S. Mondal, S., T. Paul, An integrated social network marketing metric for business-to-business SMEs. *Journal of Business Research,* **150,** 73-88 (2022)
14. O. Biernátová, Social media marketing a knihovny [Social media marketing and libraries]. *INFORUM. 17th Conference on Professional Information Resources.* Available from: https://www.inforum.cz/pdf/2011/biernatova-olga.pdf (2011)
15. J. Semrádová, *Péče o jazyk a komunikaci: (péče o logos jako smysluplnou řeč) [Care of language and communication: (care of logos as meaningful speech)].* Prague: Charles University. ISBN 978-80-7290-429-7 (2010)
16. D. Kirkpatrick, *Pod vlivem Facebooku: příběh z nitra společnosti, která spojuje svět [Under the influence of Facebook: the inside story of the company that connects the world].* 1st ed. Brno: Computer Press. ISBN 978-80-251-3573-0 (2011)
17. J. Elad, *Linkedin for dummies*. 3rd edition. Hoboken, N.J.: Wiley. For dummies. ISBN 11-188- 2221-8 (2011)
18. K. Brown, *Mastering LinkedIn*. Bookboon. ISBN 978-87-403-1064-1 (2014)
19. L. Mccarthy, D. Weldon-Siviy, *Buď pánem svého prostoru: Jak sebe a své věci chránit, když jste online [Be the master of your space: How to protect yourself and your belongings when you're online]*. Prague: CZ.NIC. ISBN 978-80-904248-6-9 (2013)
20. P. Warchar, *Jak vznikl instagram? Od nuly až k Facebooku [How did instagram come about? From scratch to Facebook].* [online]. Instagram.cz. Available from: http://www.instagram.cz/jak-vzniklinstagram-od-nuly-az-k-facebooku/362 (2015)

AUTHOR INDEX